THE ART OF THE
CRAFTSMAN

ALSO BY JESSE JAMES

American Outlaw

THE ART OF THE CRAFTSMAN

Advice, Inspiration, and
Cautionary Tales for Anyone Who
Dreams of Starting Their Own Shop

JESSE JAMES

BENBELLA

BenBella Books, Inc.
Dallas, TX

BENBELLA
BenBella Books, Inc.
8080 N. Central Expressway
Suite 1700
Dallas, TX 75206
benbellabooks.com
Send feedback to feedback@benbellabooks.com

BenBella is a federally registered trademark.

Printed in the United States of America
10 9 8 7 6 5 4 3 2 1

Library of Congress Control Number: 2025036186
ISBN 9781637747971 (hardcover)
ISBN 9781637747988 (electronic)

Editing by Rick Chillot
Copyediting by Scott Calamar
Proofreading by Denise Pangia and Marissa Wold Uhrina
Text design and composition by Aaron Edmiston
Cover design by Matt Jeffries
Cover photo by @POSTATRANDOM
Photos on pages xii and 184 by Stone; pages 100 and 128 by Eric Hameister; pages 44 and 206 by Karla James; and pages 202 and 210 by Greg Flack. All other photos courtesy of author.
Printed by Versa Press

Special discounts for bulk sales are available.
Please contact bulkorders@benbellabooks.com.

*This book is dedicated to all my shop dogs
I've had and lost for the last 36 years. They
are what I love most about my shop, and
what kept me going every single day.*

Cisco (pit bull)
Rudy (pit bull)
Noodles (pit bull)
Pee Wee (pit bull)
Milky (pit bull)
Bubba (pit bull)
Cinnabun (pit bull)
Pierre (frenchy)
Whitey (frenchy)
Duke (frenchy)
Jasper (frenchy)
Cain (pit bull)

CONTENTS

PART 3: TRUE CRAFTSMANSHIP

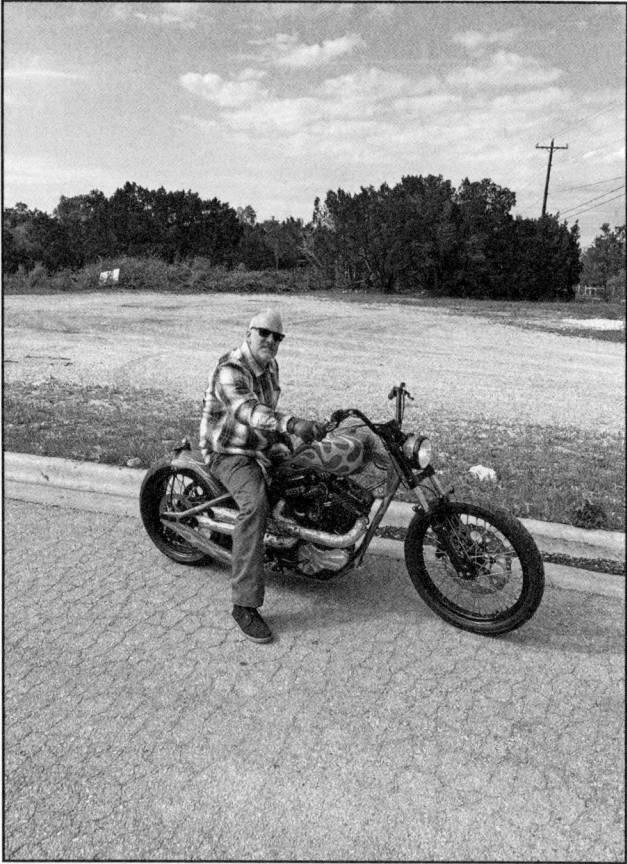

INTRODUCTION

After building a world-famous brand and multiple companies, I have one goal: helping others find success in what they love.

Throughout my career, I've been known as many things: reality star, criminal, thrill seeker, husband, father, entrepreneur. With all these titles, I still built an empire on one common theme: hard work.

Our culture is facing a mental health crisis. The era of instant gratification and "more for less" has left people without purpose, pushing them further than ever from their goals. But I know the value of hard work and how discovering your passions can springboard you to success when you cultivate and pursue them with dedication, creativity, and care.

In this book, I will share real stories of business and life to teach you how to become a better person, parent, employee, employer, and citizen of the world. Success isn't just about creating a successful business, or the fame, or the money. It's about seizing every opportunity to learn and grow. It's about the person you are behind the scenes.

Everyone wants instant gratification today. We want the success, the love, the marriage, the skills, the life, but without the hard work that comes with it. Yes, there is happiness in rest and relaxation, but we were meant to create. We're meant to work; it is an inevitable part of life. And there is satisfaction and fulfillment in putting in the hard work for the things we value.

Some people, like me, need to work every single day; more work means more happiness. One lesson I learned early on, when I officially opened my own shop, came from an unexpected place. My hours were 9 am to 5 pm. I would get up early, drink some coffee, and drive the fifteen minutes over to my shop. At least it should have been fifteen minutes. In morning rush hour traffic on the 605 and 91 freeways, it would take me at least an hour to get to the shop before 9 am. In Southern California, an hour in traffic isn't that bad. But one morning I had a customer who was passing through town early and wanted to meet at the shop at 6 am. No problem! I woke up at 4:45 in the morning and jumped on the freeway by five for my usual hour drive . . . but I was surprised that there was zero traffic. It took me only fifteen minutes to get to the shop. That was awesome, but not the lesson I learned.

When getting on the freeway at 8 am every day, I noticed most of the cars on the road were beaters: old pickups, station wagons, smoking clunkers, lowered Toyota Camrys and Nissan Maximas, the cheapest transportation possible. All typical vehicles of your average nine-to-five clock punchers just trying to make it to work on time and leave at 4:59 pm. But that one morning when I got on the freeway at 5 am—what a difference! I saw Mercedes, Porsches, shiny construction-company trucks, chromed-out concrete pumpers, nice trucks, plumbing trucks, and almost no beaters. And all of them were hauling ass at eighty miles per hour getting to work early. The visual difference was staggering.

The lesson I learned that day was very simple—and the first lesson I'll pass on to you: If you want to get shit done and accomplish your personal and business goals for success, get your ass up early and get to work before anyone else.

That will help, but it's not the only advice you'll need. So keep reading.

PART 1

GETTING STARTED

1

EMBER

I'm going to slip in some technical knowledge here and there, but this manual is actually for the most important tool in your shop: you.

It's a guide to prepare you for the mental and physical aspects of following your dreams, outlining the hard work involved in seeing that dream come to fruition. And by "fruition," I'm not talking about monetary success. I'm talking about fulfilling your self-value and knowing that you're doing the best job in your chosen discipline.

I can't be the only one who asks the question: If college professors are so smart and have all the business knowledge we need, then why don't they own and operate successful businesses?

I'm sure there are some business school professors who got their tenure due to a successful business background, but I'm willing to wager that the vast majority did not. The fact is: Working with your hands and running a business successfully is not something you can learn on an Ivy League campus. What you might learn is the fine art of sitting behind a desk as a mid-level manager with a college diploma hanging on the wall behind you.

Don't get me wrong—four years of college works for some people. However, I'm also sure that a significant percentage of those people sitting at their desks are daydreaming about following their dreams of creating, designing, and working with their hands. Very few actually pursue those dreams, and instead they trade self-value and independence for predictability and stability.

Not everything about college is bad. I spent three years at Riverside City College in California. Although I worked part-time in a shop, I was primarily interested in playing football, and RCC was the #1 JUCO football program in the country. Football was my way out. Unfortunately, I let a college counselor pick all my courses. I did my best to show up and go through the motions, but honestly, it wasn't for me. The worst example was theater class. The professor spent the entire semester on the movie/musical *West Side Story*. I was miserable. It was actually painful to sit there for over an hour listening to how great the movie, songs, and performances were. I noticed one of my teammates, who was in the same class, brought beer in his backpack and sipped it while the professor wasn't looking. After a couple of months of musical-dance-fight agony, I finally just got up in the middle of class and walked out, never to return.

My coaches swapped theater class for working in a restaurant, which I liked much better. Only later did I find out that if I had chosen my own courses at RCC, I could have taken welding, structural engineering, metal sculpture, and art. What a missed opportunity! I frequently get asked about the best way to learn to weld. I always tell people to sign up for a welding course at a local community college. It costs a couple of hundred bucks a semester, provides access to various machines, and includes all the gas, rods, and materials you need. Sometimes, you can even take a night course a few times a week. It's a very economical way to learn welding.

I feel like I'm coming at this from a very different perspective than most. Like many middle schoolers, I read *Zen and the Art of Motorcycle*

Maintenance and was discouraged to find very little information about motorcycles in the book. Don't worry—this book will have plenty of motorcycle, car, and machine information. But the real goal is to shine a huge light on boosting and maintaining your self-value while you navigate the incredibly bumpy road of following your dreams and doing what you want in life. The discipline you choose is not as important as how you execute it. I believe you'll find that much of my thirty-five years of shop experience and knowledge crosses over to many aspects of work and life. The fact that you're willing to sit, read, and learn this material shows you have the kind of rebellious spirit you need to embark on this path. It means you have that small ember burning in your gut—that vision, idea, plan, or dream. This is the best news possible for you, because without that ember, you're likely pursuing a vocation for the wrong reasons. It has to be about YOU! This may initially seem selfish, but if you have that passion, then the combination of your work ethic, integrity, skills, and knowledge will ultimately make your customers happy. This will boost your self-value and confirm that you are a success.

My life didn't go as I had planned it early

on. I could have blamed my parents, my circumstances, and my environment, and I could have
been filled with "what could have been." Instead,
I chose to focus on the things I could control and
poured my whole heart into them. Regret has no
place in a focused mind. Reaching and achieving the goals you set, whether they were part of
your original plan or they're new dreams for the
future, will give you the value you need to see in
yourself. Stop looking back and concentrate on
what makes you be you.

But here's the thing—no one else can do this
for you. That small ember burning in your gut?
It's both a blessing and a burden. It's the thing
that will drive you when everything else seems
stacked against you. You can read all the books,
watch all the videos, and attend all the classes,
but ultimately, the decisions and actions that
define your life are yours alone. This means taking ownership of your successes and your failures, learning from both, and moving forward
with renewed determination. The world is full of
people who have great ideas but lack the grit to
see them through. Don't be one of those people.

The beauty of working with your hands, of
building something from nothing, is that it's a
tangible reflection of your effort. You can look

at a completed project—a custom bike, a piece of metal art, a rebuilt engine—and say: "I did that." It's a source of pride and satisfaction that few other professions can offer. But it's also a double-edged sword. Every flaw, every mistake, every miscalculation is out there for the world to see. And that's okay. It's part of the process, part of learning, and part of becoming better at what you do.

One of the biggest lessons I've learned over the years is the importance of patience. In the beginning, I wanted everything to happen fast. I wanted to master every skill, complete every project, and achieve every goal overnight. But that's not how life works, and it's certainly not how mastery is achieved. True mastery takes time—years, even decades. It requires a commitment to continuous learning, to pushing yourself beyond your comfort zone, and to embracing the challenges that come your way. Patience is what allows you to stay the course, even when progress seems slow or nonexistent.

In the end, the most important thing is to stay true to yourself. Don't let the opinions of others dictate your path. Don't chase after someone else's version of success. Define your own goals, set your own standards, and measure your progress by your own criteria. It's your life, your dream, and your journey. Make it one that you can look back on with pride, knowing that you gave it your all and stayed true to who you are.

So go out there and create. Build the life you've always dreamed of. Embrace the challenges, learn from the setbacks, and never lose sight of that ember burning in your gut. It's your guiding light, the thing that will keep you going when everything else fails. Trust it, nurture it, and let it fuel your journey toward a life of purpose, passion, and fulfillment.

2

GENERAL INFORMATION

Okay, so what are we talking about here?

To get started on your new path, we need to gather some general information about what you want to do. Info about your goals, your dream job, and most importantly, your biggest fears about making that jump into the dirty world of busting your ass every day. Daydreaming about working with your hands in a shop is sometimes far from the reality of actually working. Giving yourself total freedom from the Aldous Huxley world of a mindless job in a mindless world can be both a blessing and a curse. It requires an incredible amount of self-discipline and even more hard physical labor, for years.

I don't care if your dream is to make dough-nuts, stained-glass windows, or custom cars. It really doesn't matter what you want to do; all of the same basic principles apply. And the first is: Start with knowledge. You need to know every single thing about what you want to do and every bit of history about it. All of this starts with books. In this day and age, with Google searches being the main source for any subject, it's hard to believe that there is still a ton (the majority) of info that's not on the internet.

My journey to building a custom motorcycle in my mom's garage started with books. Well, some books, but mostly magazines. At the same time, while doing security work for traveling bands, I was mentally building that first bike in my head. I spent a significant amount of time in Europe. And any free time I had was spent scouring newsstands, bookstores, and libraries for anything custom-motorcycle or car related. I bought hundreds of magazines, even though I couldn't read most of them because they were in other languages. I obsessed over pictures of bikes and cars that I liked, studying the lines, colors, and styles, looking at the same ones over and over again. My obsession wasn't limited to Europe. I subscribed to magazines and periodicals from

Japan and Australia. These magazines and books showed me a stark contrast from the way custom motorcycles were presented in the USA. Motorcycles and motorcycling in other countries were more about riding and craftsmanship. Gaining this visual knowledge of what other people were doing in other parts of the world was pivotal to what I built. It also still guides what I build today. And yes, I still have every one of those magazines and books.

Once I had a visual in my head of what I liked and what I wanted, the next thing needed was to learn some skills. This is probably the most important piece of the puzzle. Everyone has a business idea—it seems so easy. Just put your idea on social media, and in no time, you'll be sitting behind a big desk counting money. But it never seems to work out like that. The skills are what will carry you every step of the way. So you need to do something that you genuinely love. We've all heard the saying: "Work at the doughnut shop, and you'll hate doughnuts." Well, it's probably more accurate that you won't actually hate doughnuts—you probably will just hate *making* doughnuts. So you need to choose your discipline wisely. Physical labor, and especially physically demanding labor, can be the most fulfilling work—fun, rewarding, and a huge boost to your self-value.

Skills and processes for anything you want to do can be found in books. You are literally reading a book now about how to do something. My earliest metal and wood finishing skills were taught to me by my dad. In the 1970s, my dad had a furniture restoration business on Minnesota Ave in Paramount, California. He shared the building with Perry Sands, the founder of Performance Machine. This was the obvious early motorcycle influence in my childhood. Some of my oldest memories are of wandering next door to Perry's machine shop and playing with the metal chips made by the machines. My dad's shop was stacked high with mostly American oak furniture from the 1800s. I can remember oak press-back chairs stacked up fifteen feet high. The smell in his shop was a combination of thin liquid paint stripper, dimethyl sulfoxide, Watco oil wood stain, and lacquer thinner. The combination of these smells is permanently imprinted in my brain.

My dad had a couple of furniture auctions in Riverside and one in Paramount. He also had the contract to fill all the TGI Fridays restaurants with antiques. So my dad's shop was a busy place in the '70s. Even though I was five, six, or seven years old, my dad had no problem putting me to work. One of my jobs was to strip furniture. The furniture was completely disassembled, and the parts were placed in a big galvanized steel pan. I would brush the thinner-than-water dimethyl sulfoxide onto the painted or varnished wood and wait. Within minutes, the paint would start to bubble. I would then scrape it off with a paint scraper. Then, I would scrub all the paint out of the cracks with a stiff wire brush. You don't know real pain until you've felt stripper flicked back in your eye from one of those stiff bristles.

After all the paint was gone, I then scrubbed down the wood with coarse steel wool dipped in lacquer thinner to remove all the dimethyl sulfoxide. This was a very laborious process, and my dad would monitor my work every step of the way. If he saw a speck of paint in the wood grain, he would say things like, "Don't half-ass it," and make me finish it right. Not only did this teach me very quickly not to do things sloppily, it also showed me the satisfaction of doing a job right with hard work—taking an ugly, painted avocado-green dresser and turning it into a beautiful piece of quarter-sawn oak furniture with a warm honey-brown finish.

Little did I know that what I thought was some radical form of cruel child abuse was actually learning. The skills involved in working with my dad soon transferred to me restoring antique prewar bicycles (the first one when I was nine years old), taking something ugly and dilapidated and making it beautiful and desirable. It's an addicting feeling that I still get today, building things. I think the lesson here is that you might have some skills that you've acquired and not even realize it. This is the general information that we need to flesh out to get you on the path of doing what you really love.

Also, don't think I'm telling you that you need to be aware of these lessons happening in your life, and don't think that you have to have these kinds of lessons at all. The reason I'm telling you these stories is to give you some perspective on my mindset on these topics. The fact is, I didn't realize I even learned anything when I was younger until way later in life. Although you do need something inside—starting with that ember we talked about last chapter—to give you a push.

It's not about where you are currently; it's where you want to go. If you don't know, find out. If you can't do, learn. If you don't have the heart to take failure, don't start. Don't let your doubts, lack of education, or pride get in the way of doing something that you love. Education has become this weird word in the world today; I'm not talking about education in college or graduate school (which all serve a purpose and give something in return), I'm talking about educating yourself with the resources you have at your disposal. Read, visualize, execute.

Success isn't always about the end product. Sure, it's satisfying to look at a finished project and feel that sense of accomplishment. But the real value lies in the process—the long hours, the problem-solving, the trial and error, and yes, even the failures. Each step you take, each mistake you correct, and each lesson you learn builds not only your skills but also your character. It molds you into someone who doesn't just talk about dreams but actually lives them.

Remember that ember we talked about? That small, persistent drive inside you? It's not enough to just let it burn quietly. You've got to feed it, stoke it, and turn it into a raging fire that propels you forward, even when things get tough. Because make no mistake, they will get tough. There will be days when nothing seems to go right, when you feel like you're spinning your wheels, or when you question whether it's all worth it. Those are the moments that define you. Will you let the fire die down, or will you use those challenges as fuel to push even harder?

At the end of the day, your journey is your own. No one else can walk it for you, and no one else can define what success looks like for you. You're the one who has to live with the choices you make, the work you produce, and the life you build. So make it one that you're proud of. Make it one that reflects your values, your passions, and your commitment to excellence.

3

THE LEAP

Just getting started is probably the biggest initial obstacle you will encounter. This obstacle is all based on fear. If you're working a full-time job and making a decent paycheck, the idea of quitting and becoming a craftsman seems crazy. Not only does it seem crazy to you, but everyone around you will likely think you're an idiot. This is because most people base success on monetary value. Quitting something that sucks and moving on to something else for more money—that's an idea society will cosign. But doing it with no financial benefit, no guaranteed success, and no safety net? Everyone is going to treat you like you are terminally ill.

You yourself may even question this choice. In 1995, I quit working for Hot Rods by Boyd and moved West Coast Choppers from my garage into my own shop. It was a 3,000-square-foot building rented to me by hot rod legend Doyle Gammell. The Minnesota Ave shop in Paramount was also a former location for Performance Machine and for my dad's shop. And it was directly across the street from the first building they shared. My first week in business, working solely for myself, was pretty nerve-wracking. Suddenly, I had no paycheck, but I had rent and utilities. My head was clogged with every possible scenario of how this could all go wrong.

After a couple of days in the building, my friends Fast Eddie and Jim Lilligard came by to see how the shop was. We chitchatted for a minute, and then I continued to work on a bike I was building. Jim and Fast Eddie just walked around the shop, looking at what I had going on. I saw Jim lean over to Fast Eddie and heard him say, "He'll be out of business in a month." Now, whenever I recall this story, I always think *Fuck those motherfuckers! I've made it thirty years!!* But back in 1995, when I overheard it, all I felt was fear. I didn't need their verbal support, but it struck me that nobody believed in my idea or vision. This just fueled my fire to succeed.

If you're really going to do this, you'd better get used to not having anyone support or even see your vision. It's really not that big of a deal because if all these negative assholes were able to see what you see, they'd all try to do the same thing.

I have some practical advice that will help. As far as getting a building and creating overhead, I waited until I was bursting at the seams. I had more work than I could fit in my two-car garage, and I had a backlog of orders that I could no longer fill by just working nights after my day job. I highly suggest you do the same. Timing is everything, and waiting is the safest way to approach it.

In my new building, I was scared for about two weeks. Then I got so buried with bike and parts orders that, to this day, I've never been able to get completely caught up. There's always a certain security in having a backlog.

I would also suggest keeping your overhead as low as possible. It's very easy for it to balloon up quickly. As soon as you get a business license, the floodgates of profit vampires will open up. Sales guys will start stopping by, trying to sign you up for uniform service, floor mats, shop rags, etc. They make it sound so easy and like the smart thing to do. Before you know it, you'll be spending $700 a month for mats and shop rags. No thanks. I went to Costco and bought my own mats and rags. I also found a wringer washer in *The Recycler* for $200. Wringer washers are cool because you can put a load of shop rags in them with some Tide and let the washer run all day. The rags come out like brand new, and there's no dumb $700-a-month mat/rag service bill. The leaner and meaner you can do things from the very start, the easier things will be down the road.

Follow your dream and do something you love. It's really like jumping off a cliff—an all-or-nothing approach. Jump-start yourself into action. Don't be afraid to take the leap. Society will tell you "no," "stop," or "bad idea," but you know yourself better than they do.

Every big step I've taken has scared me—if it were easy, everyone would do it. The first step is the hardest, but after that, you just have to tighten your belt and get to work. Throughout my entire life, I've found that keeping my overhead low allows me to live freely, turn down customers I don't want to work with, pay my people well, and do what I want to do. Giving yourself permission to dream, succeed, and fail will define you as a person. Sometimes, you just have to take that step to see what you're made of.

NO PERFECT TIME TO START

I've talked to thousands of people who've started their own businesses or trades, and the one thing they all have in common is this: It wasn't planned. There's no perfect moment to say "On the fifteenth, I'm quitting my job, and on the seventeenth, I'm starting my new life." That's not how it works. It happens organically. Most people are working a boring, unfulfilling job when they realize they've got something better going on part-time. They're in their garage after hours, working on a project, and it hits them: *This is what I want to do. This makes me happy.* At that moment, they take a leap of faith. They might make less money at first, but they're more fulfilled, happier, and they're living on their terms.

And that's the key—living on your terms. When you work for someone else, you're building their dream, not yours. You're helping them achieve their goals while you sit back and wait for your paycheck. But when you take that leap, when you decide to go for it on your own, you're building something that's uniquely yours. It's hard, it's risky, and it's not always glamorous, but it's real. And that's what makes it so fulfilling.

When you take that leap, it's scary. You're walking into the unknown, and there's no certainty. But here's the thing: If you're not willing to take that risk, you'll never find out what you're truly capable of. So many people sit on the sidelines, waiting for the perfect time, the perfect circumstances, the perfect opportunity. But the truth is, there's no such thing. The only perfect moment is the one you create. The hardest part is making that decision—deciding to go for it even when you don't have all the answers. You can sit and plan for years, but until you take action, you'll never know what's on the other side of that leap. And trust me, it's worth it.

It's not about following some rigid business plan or financial road map. It's about listening to your gut, following your skills, and letting your heart guide you. If you've been given a gift—a talent, a skill, or a craft—you owe it to yourself to use it. It's like God reaching down and saying, "Now's the time. Go do it." That's not something you can schedule. It's something you feel. And when you feel it, there's no holding back. The passion drives you forward, even when you don't have all the details figured out. You figure it out as you go.

The first time I took that leap, I didn't have a solid plan. I didn't have a five-year road map laid out. What I had was a passion, a craft that I couldn't ignore. And once you get a taste of that fulfillment, there's no going back. It's the difference between living for the weekend and waking up every morning with a sense of purpose. People often ask me how I built West Coast Choppers, how I turned it into a business that people love. The truth is, I didn't focus on the business side of it first. I focused on the craft. I made sure I was doing something I was proud of. The rest—success, money, fame—all of that came later.

When you're doing something that lights you up, when you're working for yourself, everything else falls into place. It's not about chasing money—it's about chasing fulfillment. When you're truly invested in your craft, the rest of the world will take notice. The customers, the fans, the collaborators—they see that energy, that authenticity. And they'll be drawn to it. So don't wait for the "perfect time." The perfect time is when you're ready to go for it. The rest will follow.

THE LEAP INTO BUSINESS

It wasn't easy to leave a steady paycheck. Going out on your own is terrifying because there are no guarantees. I rented a shop, and for a week and a half, I was scared out of my mind. But then the work poured in. Fax machines spit out purchase orders nonstop, bikes rolled into the shop, and I was drowning in work. I've never looked back. From 1993 until now, I've never "caught up." There's always another job, another project, another deadline.

But even with the success, I've never lost sight of the value of hard work. The only reason I'm where I am is because I dedicated myself to the craft. I kept learning, kept pushing myself, and kept improving. The world needs more people who are willing to do the same. It's not about getting rich or famous—it's about finding fulfillment in the work itself, in the pride that comes from creating something with your own hands.

RUNNING A SHOP: ECONOMICS AND MINDSET

When I built my business, I never took loans or lines of credit. If I needed a new lathe, I'd figure out how many fenders I needed to sell to pay for it, then I'd buy it outright. That's how I grew my shop—one tool, one project at a time. Most people think you need debt to start a business. That's a mistake. Debt will bury you. The moment you can't pay, you lose everything. I built everything I have—CNC machines, tools, the shop itself—by reinvesting every dollar I made back into the business. If you approach it with discipline, focus, and respect for the craft, it's possible to create something truly special.

In the end, it's all about respect: respecting the work, respecting the tools, and respecting the people who do it. There's nothing more important than that.

WHEN TO TAKE THE LEAP

So how do you know when it's time to start? You need the skills first, yes, but waiting for the "perfect" time is a trap. It's never going to feel safe. You'll never be 100 percent ready. At some point, you just have to take the leap. If you have the passion and the skills, the rest will follow. But that leap? That's where most people freeze. Fear keeps you from making the move. You start questioning yourself: Am I good enough? What if it doesn't work? What if I fail? Let me tell you, the "what-ifs" are the lies that stop most people from even trying. If you're waiting for the right time, it's never gonna happen.

When I decided to quit working at Boyd's and start West Coast Choppers as my sole source of income, I was terrified. I had everything set up—my business was thriving, I was in catalogs, my manufacturing process was dialed in, and I was running things the right way. But even with all that, I was scared. *What if this takes a shit? What if the business dries up overnight?* I worried about losing everything. I pictured having to swallow my pride, go back to a regular job, and ask for work. That fear of failure—that fear of tucking tail and admitting "It didn't work out"—that's what stops most people dead in their tracks before they even start.

But here's the truth: There's never going to be a "right" time to make the leap. Life doesn't come with a perfect moment or a guarantee that everything will work out. You have to jump anyway. At some point, you have to trust yourself and take the plunge. You have to follow your gut, follow your heart, and know that what you're doing is what you're meant to be doing. If it's not authentic—if it's not something you love—you're not going to make it. It can't just be about the money. If you're copying someone else or chasing what's trendy, you're setting yourself up to fail. Why? Because the only way someone's going to buy a copycat product is if you're cheaper than the next guy. And that's not a path to success. That's a race to the bottom. When it's about following your passion, everything else becomes secondary. You'll outwork anyone who's doing it for the wrong reasons.

IF IT'S FROM THE HEART, YOU'LL SUCCEED

When you're doing something you truly love— something that comes from the heart—nothing will stop you. It's not work; it's who you are. That's what sets apart the people who succeed from the ones who quit. When it's your passion, you'll work harder than you ever thought possible. You'll put in late nights, early mornings, and sixteen-hour days without even realizing it because you care. It's your heart, your soul, and your craft.

I've seen so many people go into a trade or a business thinking, *This is where the money's at.* Maybe they're a great welder, but they decide, *Screw it, I'll make cabinets instead. I'll buy a router, some wood, and a planer. How hard can it be?* At first, it might seem fine—they're making decent money. But eventually, it turns into a grind. They don't care about the work. They don't love it. Pretty soon, they're cutting corners and dreading every job. That's when it all falls apart. When your work isn't authentic to who you are, it shows. The people who truly succeed are the ones who love what they're doing. You can feel the difference when you see their work. You can tell it's from the heart. And when your work comes from the heart, people notice. They trust you. They respect you. And they come back—over and over again.

THE LONG ROAD AHEAD

Taking that leap isn't easy. There's no manual to tell you when it's the right time. But if you're driven, if you're passionate, and if you have the right skills, trust that you'll figure it out. The road ahead might be long, and it might be tough, but if you have the grit and the heart, you'll make it. Don't wait for the "perfect moment" to start. You'll never get it. Take the leap, build something that matters, and pour yourself into the work. Because in the end, that's the only thing that'll get you through the hard days and lead you to something worth having.

When I first decided to start my business, I had no guarantees. I didn't know whether it would work or if I'd make it. But that uncertainty is what drove me. The unknown is always a scary place, but it's also the place where growth happens. Sure, there's risk involved—but without risk, there's no reward. The truth is, you won't find success sitting on the sidelines. You have to jump in, make your mistakes, and learn from them. The fear of failure is real, but it's also a sign that you're pushing yourself toward something bigger.

PART 2

KEEPING IT GOING

4

ENGINE

The 1957 Cadillac was king. Not only in America, but worldwide. The top-of-the-line 1957 Cadillac Brougham cost $13,074 dollars. For comparison, a same era Rolls-Royce cost $10,000. American led the way on many fronts. The engine powering the 1957 Cadillac was a 365 ci V8 with 10:1 compression ratio and two four-barrel carburetors, putting out a conservative 325 horsepower. Sadly the Rolls only put out 180 horsepower. A big luxurious car needs big luxurious horsepower. Having plenty of power at your disposal is always going to put things at an advantage.

So far we've narrowed down our ideas and covered the ways to gather inspiration for these ideas. Now it's time to make that move, to work on what we want to be doing. Your new business is going to have a lot of moving pieces, and just like the Cadillac, it's going to need an engine with some substantial power to make it all work. News flash! The engine is YOU. Most people think a building, and a bank loan, and a partner are all you need to make a successful business. But my feelings differ. I believe the most important thing is you—the engine. You are the thing that is going to move everything forward and keep it moving forward. That engine needs to be strong because you're going to work your ass off for the next seven years to make this thing stick. Weekends, holidays, birthdays, anniversaries. You're going to work straight through all of it. The good news is you don't have to pay yourself time-and-a-half overtime. So you're already saving money! But it goes without saying that with the amount of hours/weeks/months/years you are going to have to bust ass, you better really pick something that you love to do. Because that's all you're going to be doing.

Now that I've made it clear that a lot of hard work is going to be involved, we should probably discuss how you're going to sustain this hard work. West Coast Choppers has been in business for thirty-five years. For the first couple of those years, I was working a full-time job for someone else and running my motorcycle-building business at night out of my garage. I would finish my day job and rush home and go back to work in the garage. The key trick in this situation is not to go in the house, and for sure do not sit on the couch, not even for a minute. The couch has a very strong gravitational pull on your ass. Once you get caught, it is almost impossible to get away. So when it's time to start working, go straight to the garage, shop, or wherever it is you're doing your thing.

In my two-car garage on Hackett Ave in Long Beach, I had just about everything I needed. I had a Clausing 8530 baby mill, a Chinese lathe, an air compressor, a Miller Dialarc 250 welder, and a paint booth made from PVC pipe and Visqueen that was on pulleys. I could pull a rope, and the booth would raise up into the rafters and out of the way. Also, hand tools, polishers, and everything needed to color and polish paint jobs. In fact, I had so much junk piled into that two-car garage, it would take me half an hour every day just to move everything out so I could get to work.

In my mid-twenties, my engine was strong. Working two full-time jobs building custom cars during the day, and painting and building bikes at night, I had the world by the nut sack. But after several years of burning it at both ends, my engine started to experience some problems. It took me about twenty years to realize what "balance" was. You can only push yourself so far without taking care of yourself and doing proper maintenance. This includes living a healthy lifestyle: getting real rest, and putting the right fuel in your engine (i.e., your diet). Doing my TV series *Monster Garage* gave me a very good example of how to pace yourself. Most builders would come on that show all gung ho. They would work till the wee hours of the morning during the first two days. Then by the third day, they were useless and already burned out. I would take a more relaxed pace and work ten-hour days Monday through Thursday. Then on Friday when we had a 12 am deadline, I would have some reserve in my tank to bust ass and finish the build by midnight. That same *Monster Garage* method of pacing myself applies to the way I work today. It's simple: Do good work and make substantial progress on all of the workdays.

Working when you're tired is something that will have to happen, but you need to listen to your body. You will also need to control your diet and your alcohol consumption. I gave up drinking twenty-four years ago purely because I could not have it fucking up my next workday with a hangover. So with alcohol and other recreation, you need to decide what is more important. I'm not saying to be successful you need to be sober. But I do know plenty of people who chose partying and gave up on seeing where their skills could take them.

Now for diet, it's pretty simple to understand. If you're a physically healthy person, you're going to be able to put in a healthy amount of work. You will also be able to sustain working longer and harder. Fat slobs are always going to peter out after their Del Taco lunch of a burrito and a large Coke. But the guy who has just the right amount of protein and plenty of water, he's working hard all day and going home and straight back to work in his shop. Case in point: Gene Winfield was on *Monster Garage* when he was almost eighty years old. He was part of the Dream Team that chopped and channeled a 1954 Chevy. Gene completely busted his ass that week working on the car. He was all over it. Chopped the top and hammer welded all the welds. Helped Dick Dean channel the car four inches. Both are big jobs to try and pull off in a five-day week. But he did it, and he painted the car a classic "Winfield fade" when it was done. How does an eighty-year-old do this? Gene was on a strict high-protein, low-carb diet. He prepared all of his own meals, and every couple hours he would stop and eat. I took notice of this because I want to still be kicking ass when I'm eighty. A healthy, powerful engine needs good fuel and proper maintenance to last a long time.

Your body is your engine. How you maintain it and how you fuel it directly affect your performance—not just at work or home but wherever you are, both mentally and physically. Many problems can be resolved by getting into the weight room, running on the treadmill, and eating clean. Too often, people neglect to pay themselves first. This means you have to invest in your body. You must prioritize putting energy into yourself before anything else. If you don't, by the time you finally have the opportunity, you might find yourself completely depleted and losing ground. It may not happen overnight, but after consistently neglecting yourself and failing to invest in your well-being, you'll look back years later, facing poor physical health and mental health issues, wondering how you got to this point. Pay yourself first. Get up and work out.

CAR CULTURE LEGEND, GENE WINFIELD

Robert Eugene Winfield (June 16, 1927–March 4, 2025) was a legendary American custom-car builder, painter, designer, and dry-lakes racer. Known as "Windy" among friends, Gene began restyling cars in a chicken coop behind his mother's house at 1309 Figaro Avenue in Modesto, California, after WWII. Seventy-three years later, he was still running Winfield's Custom Shop. The shop had been relocated to Mojave, California, and Gene was touring all over the world chopping cars and laying down his signature fadeaway paint jobs. He was inducted into Darryl Starbird's National Rod and Custom Hall of Fame, and was honored as "Builder of the Year" at the 2008 Detroit Autorama.

LOVING THE WORK

If you love what you do—if you love your shop, your craft, and the fulfillment of creating something with your own hands—you won't have to force yourself to work hard. For me, the struggle is the opposite: I have to stop myself from working too much. I love the process so much that I have to consciously tell myself to take breaks, to leave the shop at night, to give myself time to rest. The drive to create, to make something new and beautiful, can be all consuming. But I've learned that the work doesn't end when the shop door closes. Taking time to rest is part of the cycle, and if you ignore it, you'll burn out fast.

Every day starts early for me. I go to the gym, I shower, and I'm in the shop as early as I can be. The shop is where I feel alive, where I connect with my craft. The focus and energy are unmatched. By the afternoon, I give myself a short-term goal: "Okay, I'm going to finish this one last thing, and then I'm done for the day." It's hard to walk away sometimes, but discipline applies to rest too. Without it, you'll burn out. There are days when I feel like I could keep working through the night, just pushing forward on a project, fine-tuning every detail. But I've learned that the quality of the work will suffer if I don't take that time to rest and recharge.

And rest doesn't just mean taking a break from the physical labor—it means giving your mind the space to reset. Sometimes, when you've been working on a project for hours, you become too close to it. You stop seeing the mistakes. Your mind starts running in circles, focusing on the wrong things. That's when stepping away is the best thing you can do. It's not about giving up; it's about taking a step back so you can come back with fresh eyes and a renewed focus. It's a lesson I've had to learn the hard way: Working through exhaustion doesn't make you more productive. It makes you prone to mistakes that will waste time and cost you more energy in the long run.

In the end, it's all about discipline. It's about efficiency, focus, and a relentless dedication to the work. Whether you're building motorcycles, forging knives, or crafting heirloom cookware, success comes down to showing up, doing the work, and never making excuses. That's the foundation—showing up every day and putting in the effort, regardless of how tired or frustrated you might feel. But there's a balance. Pushing too hard, working through exhaustion, will only lead to mistakes. And that's a lesson I had to learn the hard way.

RECOGNIZING WHEN TO STEP AWAY

You know, I go through the same things that everyone does. I get tired. I get burned out. I get frustrated—sometimes it feels like no matter how hard I try, the more I touch a project, the more I screw it up. And that's when I've learned one of the most important lessons: You have to know when to stop. When the frustration builds, when I can feel my hands losing finesse, when I start forcing it instead of letting the work flow naturally, I know it's time to step away. I'll tell myself, "Okay, I'm done for now." I'll go eat dinner, get some rest, and come back to it in the morning.

Well, I think it's kind of my version of writer's block. Like how a writer just can't get words on paper, when you're building something from scratch—creating from raw material—the block can feel even more intense. It's three-dimensional. It's not just in your head; it's in your hands.

There are two versions of it. One is where you physically can't get started. You can't pick up the tools, can't bring yourself to begin. The other is when you are doing something, but what's coming out just looks like shit—substandard, not right. And that's actually a good sign, in a way, because it means your standards are high. You're on the path to being a craftsman. You have a vision and you're not willing to accept less than that.

But you also have to know when to stop.

Forging Damascus (a type of pattern-welded, layered steel), for example—if something goes wrong, like the metal doesn't fuse or it fractures or overheats, you have to stop. When it's at 2,100 degrees, it's not coming back. It's only going to get worse. You learn to recognize that moment: Okay, this broke. Time to let it cool down. Step away.

Same goes for sheet-metal fabrication. You can be forming a panel, hammering and hammering, and it just keeps getting worse. You reach a crossroads: If I keep going, am I making it better—or just making it worse?

That's a decision you've got to make. Sometimes, you need twenty-four hours away from the thing. And sometimes, you come back and realize: This one's done. Not in a good way. Trash it. Start over.

That's a hard thing to do. Most people will finish a project and try to justify it based on the time they put in. But I've never been willing to do that. I've tried a few times, and every time I end up going, "Screw it. If it's dog shit, it's dog shit." Doesn't matter if I spent 10,000 hours on it—if it's not good, it goes in the trash.

And that moment—that threshold—is brutal. Especially for someone who works with their hands. Because we want our skills, our work ethic, our discipline, to mean something. We want them to produce something real. And sometimes, they just don't. Sometimes you spend days, weeks, months . . . and get nothing.

But you have to treat that as part of the process. As a lesson. That's how you learn what not to do. Maybe this time you welded too much—or not enough. Maybe you skipped a step, or you didn't do it right. And the truth is, any corners you cut in the beginning will come back to haunt you at the end. Every single time.

Almost every single time I do that, I wake up with a fresh perspective. It's amazing what a night of rest can do—suddenly, I see the problem differently. My mind is clearer, my approach is calmer, and I can focus on the details I was missing the day before. I've learned that pushing through when I'm mentally or physically drained only leads to mistakes—breaking something, busting a tap, ruining hours of progress. That's when it costs you time and energy. You don't want to get to that point. The real skill is learning when to walk away before that happens.

The key is recognizing when you're not doing your best work. It's a discipline you have to develop. You've got to step back before you make a costly mistake. And it's hard because we're all stubborn—especially when you're working for yourself, building something you're passionate about. You want to finish. You want to solve the problem. But sometimes, stopping and letting the project breathe is the best thing you can do. It's not a failure to take a break; it's an investment in the quality of your work. It's saying, "I care enough to do it right, even if it means stepping away and coming back with fresh eyes."

When you love what you do, it becomes an obsession. I don't just "work" on projects—I dive into them. I lose myself in the process. It's not about rushing to the finish line; it's about the satisfaction of the journey. That's when you do your best work. When you're enjoying the work itself, it doesn't feel like a job. It feels like a passion, a calling. You're not just building products; you're building your legacy, your story, and your satisfaction. The work itself becomes the reward.

THE BALANCE OF REST AND DRIVE

We talked about this idea of knowing when to stop, of taking breaks. And I'll admit, it's something I struggle with. Over thirty-five years, I think I've taken maybe four vacations. I'm not a big "vacation guy." Skiing? Forget it—I'd probably just mess myself up. I love the beach and the water, but the hardest part for me is shutting my brain off. Even when I'm not in the shop, my mind is still there, thinking about projects, ideas, and all the things I haven't built yet. The work never really leaves me, and in some ways, that's both a blessing and a curse.

The longest that I've ever stepped away from the shop was for two months—and even then, I wasn't really resting. I was managing things, organizing, still working but in a different way. But what I've learned is that even if I'm not good at resting, it's necessary. You have to give your body time to recover. You have to let your mind reset. Otherwise, you're just grinding yourself down, and the quality of your work will suffer. It's like trying to run a car on an empty tank—you might be able to get it moving for a little while, but eventually, it's going to stall. And if you keep pushing, you're going to burn out.

The flip side, though, is this: I've got so many ideas. I've got so many things I want to make, to try, to build. My biggest fear in life is dying before I've built all of it. That's what keeps my foot on the gas pedal. I don't want to sit by a pool or lounge on a beach for days at a time, because every minute I'm not working feels like a minute wasted. I have projects in my head, sketches on paper, and ideas brewing at all times. I can't help it—I love creating. But like I said, I know that the drive and ambition can't come at the cost of my well-being. So it's an ongoing battle to find the right balance between pushing forward and stepping back to recharge.

And maybe that's not for everyone. Maybe you're at a point where sitting on a beach is better than your day job. But for me? I love what I do. I don't need an escape. In fact, my work is my escape. There's a real sense of fulfillment in it. When you're working on something with your hands, solving problems and overcoming challenges, it's like you're in the zone. Time doesn't exist. It's just you and the work. The satisfaction that comes with building something from nothing is like nothing else. The idea of taking a long break from that is hard for me to understand, but I recognize that I can't keep going without those breaks.

That's why, over the years, I've had to learn the importance of rest—not just physically but mentally too. Rest is more than just sleep. It's about giving your mind the space to breathe, to let go of the constant cycle of problem-solving. Because when you're constantly thinking about the next project, the next idea, it's easy to overlook the details in front of you. You lose perspective, and that's when the quality starts to slip. The thing is, you can't create good work if you're always running on empty. The best ideas come when you've had the time to let them marinate in the back of your mind. Sometimes, stepping away from the shop is what allows the best ideas to surface.

Remember, the seed you planted won't grow on its own. You've got to water it, nurture it, and give it time. But don't forget to rest the soil too. Without proper care and balance, even the strongest seed can wither.

5

POWER STEERING

Weighing two and a half tons and measuring eighteen feet long, the enormous 1957 Cadillac would be virtually undrivable without power-assisted steering. A belt-driven hydraulic pump on the engine powers a steering gearbox. The gearbox is essentially a piston with a gear cog. Fluid pressure, controlled by the steering wheel, moves that piston and cog back and forth, turning a gear on the output shaft connected to the pitman arm. This, in turn, moves the wheels left or right. Without that pump and fluid pressure, the whole mechanism is almost immovable. You could get it to move, but would be like hand-over-hand rope climbing to get the steering wheel to turn even a small amount.

In choosing the life of a craftsman and committing to making things in a great way, you are the power-steering pump. Without your pressure and hard work, nothing moves easily. You are the one who has to do all the dirty work—from building things, to ordering materials, to dealing with customers and taking their money. You need to do it all. When the shop toilet clogs, you are the one who fixes it and cleans up the mess. As I mentioned earlier, you need to love what you are doing so much that you are willing to handle all of this without procrastinating. And in fact, all of these tasks are really a blessing to manage. Waiting around for someone else to handle things will only lead to frustration. And at the end of the day, nobody will handle things as well as you. So my advice is to do everything now, and don't wait. Don't ever say "I'll do that later" or "tomorrow," because tomorrow will bring a whole new list of things that need to be handled.

That list will all be the smallest, seemingly unimportant stuff, like walking by a piece of trash and not picking it up, not refilling a cutting oil bottle, not rolling up the air hose, or not draining the water out of the air compressor. Eventually, you'll have to do all of this, so do it now. Burning up the motor on the air compressor because you were too lazy to drain it once a month will only cost you $7,000 to buy a new one. This is mostly shop maintenance stuff I'm talking about, but this discipline also applies to your work and your business practices. Calling someone back or responding to a text also needs to be handled now. I find that if I do everything now, the tasks I need to handle don't pile up as much as they would if I waited. I'm also a very impatient person, so when I'm in the middle of a project and I need parts or materials, I cut the time it takes to get stuff or information by handling it immediately.

There is nothing more frustrating than being in the middle of something, hot and heavy, trying to get it done, and having to wait on a part or some key element. When the person you're waiting for takes their sweet time . . . that makes me want to reach through the phone and strangle someone. This is also a big reason why I do so much myself—the only person I'm waiting on is me. I'm the power-steering pump. When that steering wheel turns, the pressure is on.

The massive 1957 Cadillac, with its hefty weight and size, relies heavily on power-assisted steering to remain manageable. The hydraulic pump, driven by the engine, is the unsung hero of the steering system, transforming the steering wheel's input into effortless wheel movement. This intricate mechanism underscores the importance of the power-steering pump—without it, maneuvering such a behemoth would be nearly impossible.

You are the power-steering pump. Your dedication and effort are what make the operations smooth and manageable. Embracing every task, from the grand to the mundane, is crucial. The meticulous attention to even the smallest details, like fixing a clogged toilet or refilling an oil bottle, prevents larger, costlier problems down the line. Procrastination only adds to the pile, making the load heavier and more frustrating.

So tackle tasks promptly. By handling responsibilities immediately—whether it's responding to calls, maintaining equipment, or managing customer needs—you keep the wheels of your craft and business turning smoothly. Waiting only leads to inefficiencies and frustration. Being proactive and taking ownership of every aspect of your work not only streamlines your processes but also ensures that you are the master of your domain, with no one else to blame or wait on. Just like the power-steering pump, your commitment is what drives the success and efficiency of your craft.

6

THE IMPORTANCE OF DISCIPLINE

None of the hard work we've been talking about happens without discipline. When you're your own boss, there's no manager standing over your shoulder telling you what to do. You don't have a shop foreman barking orders, reminding you to clean up or finish a job. You are the boss. You're the manager, the foreman, and the driving force behind every dollar you earn. That requires an almost militaristic level of self-discipline.

Discipline starts with the little things. You drop something on the floor? Pick it up—right now. If you finish a task, clean up your tools immediately. If you get an email, a text, or a call that needs your attention, handle it right now.

I don't make lists. I don't schedule tasks for tomorrow if I can do them today. That's the difference between efficiency and procrastination. Putting something off, no matter how small, will only slow you down later. Eventually, you'll have to do it anyway, and chances are it'll come up at an inconvenient time when your workload has already snowballed. Discipline is about taking care of things immediately, staying on top of everything, and making sure nothing slips through the cracks.

I've seen it a million times—guys who procrastinate, guys who keep messy shops, who don't maintain their tools, and who don't communicate well. All of those bad habits roll over into the rest of their business. They miss deadlines. They don't market themselves well. They can't meet the demands of customers. And eventually, they're not making much money or achieving any real success—they're just treading water. Inefficiency kills craftsmanship and kills business. If you don't have the discipline to stay organized, take care of your tools, and communicate well, you'll struggle to keep your business afloat. In the end, it's the little things—the attention to detail—that make all the difference.

You ever seen someone ignore something small—something dumb like a screw on the floor—and next thing you know, it turns into a huge problem? Happens fast. Maybe someone steps on it. Maybe you roll a bike tire over it and now you've got a flat. All because somebody didn't take two seconds to pick it up.

That's where the "do it now" mindset comes in. That principle—work ethic, whatever you want to call it—comes to me straight from my dad.

He was relentless. From the time I was seven or eight, it was always "do it again, do it better." One of his most-used phrases was: "Don't half-ass it." And if I did? He'd make me redo it until it was right—even if I was crying, even if all I wanted to do was go play outside. I didn't want to be scrubbing paint stripper off oak furniture with steel wool and bare hands. I hated it. I hated him for it.

But now? I'm thankful.

Because that work ethic is baked into me. It's automatic. He risked me resenting him—and maybe he didn't even realize it at the time—but he chose to teach me the big-picture stuff: doing things right, not being lazy, having pride in my work.

It's a hard line for parents to walk. I couldn't do the same with my own kids. I wanted them to have a better, easier childhood than I did, but looking back, maybe I did them a disservice by not pushing them harder. I didn't realize how much good my father's toughness did for me until much later in life. His discipline set me on a path I might not have found otherwise.

And yeah, there's a downside. I have a hard time relaxing. I feel guilty taking a day off. I hear his voice in my head every morning, yelling at me to get up. Literally throwing ice water on me to get me out of bed. Stuff that would probably get CPS called these days. But it worked.

Now, I don't do the work for him. I do it for my business. For my shop. And I carry that mindset with me into everything I build.

Because how you keep your shop—how clean, organized, and dialed in it is—is a direct reflection of the work you produce. If your shop looks like a mess, your product's gonna carry that same energy. Chaos in, chaos out.

Clean shop. Clear head. Better work.

NO EXCUSES: THE MYTH OF "TOOLS AND MONEY"

A lot of people will read this and say, "Well, he's famous and rich. Of course, he can do this." I've heard it all:

"If I had those tools, I could do it too."

"If I had that kind of money, I'd be successful too."

Here's the truth: Those excuses are a smokescreen. If you don't know how to use the tools, they're worthless. If you had my shop, you wouldn't even know how to turn on some of the machines. Tools don't build success. Work builds success. The work pays for the tools. The work maintains the tools. And the work is what ultimately makes you valuable.

People always have excuses. I've seen it my whole life. They'll say, "My car broke down, so I couldn't get to the shop on time." They'll show up thirty minutes late to their own business, every single day, not realizing the math of what they're losing. Let's break it down:

Showing up thirty minutes late every day equals 2.5 hours a week.

If your shop charges $200 an hour for labor,

that's $500 a week you're losing because you were lazy.

The math doesn't lie. Every minute you're not working, you're giving money away. Stop making excuses and start putting in the work. That's the only real formula for success. The moment you stop blaming the tools, the money, or any other external factor, you're free to focus on the one thing that actually matters—the work.

Excuses are like quicksand. They suck you in and keep you stuck. It's always easier to point at an external reason why things aren't going the way you want. But if you want to succeed, you need to own your circumstances. Your shop, your tools, and your resources may not be perfect—but your work ethic, your ability to problem solve, and your discipline can make up for it. So stop worrying about what you don't have. Work with what you do have. And if you don't have the right tools yet, start building the discipline to earn them.

WORK ETHIC: SIMPLE MATH

People have this romantic idea of owning their own business. They think it means sitting back in a fancy office, feet up on the desk, taking calls and calling the shots. That's not how it works. If you're the owner, you're not just a manager—you're the hardest worker in the building. You're the one staying late, solving problems, and making sure the job gets done right. The truth is, running a business is a grind, especially in the beginning. But it's not about complaining—it's about working smarter and harder than anyone else.

A long time ago, someone asked me why I was successful. My answer was simple: math. I work twice as many hours as everyone else, so I make twice as much. That's all it is. There's no secret formula, no shortcuts. You have to outwork everyone else. When you're running your own shop, there's no overtime pay. Every extra hour you put in is yours. And that's how you build something real—something lasting. You work harder, longer, and smarter. The people who fail are the ones who think they can half-ass it. They think running a business means working less, not more. But here's the thing: The first few years of any business will be the hardest years of your life. You're going to have to pour every ounce of energy, time, and money you have into it. You won't get a break. You won't take vacations. You won't sleep much. And you definitely won't be clocking out at 5 pm. But if you love it—if it's yours—it won't feel like a sacrifice. It'll feel like building a legacy. And when you're doing work you're proud of, you'll want to put in the extra hours. You'll want to go above and beyond, not because you have to, but because it's who you are.

A REGIMENTED ROUTINE: DISCIPLINE IS EVERYTHING

At this point in my life, I know I do my best work—and my best thinking—when I follow a disciplined schedule. I get up at 5:30 am every day. By 7 am, I'm in the gym. I work out for an hour or so, shower, and I'm in the shop by a little after 9 am. I work straight through until 5:30 or 6 pm every day. That routine is my anchor. It keeps me clearheaded, focused, and productive.

When you're operating at a high level, your body and mind need to be in top shape. You can't afford to show up sluggish, mentally checked out, or unhealthy. That's when mistakes happen—and in my world, mistakes can cost a lot. Every day that you're in the shop, you're working with machines, tools, and materials that are unforgiving. If you're distracted, if you're tired or not focused, that's when accidents happen. It could be something small, like a slip of the hand that ruins a piece of work—or something more serious, like a mistake that puts your entire team at risk. That's why discipline matters. When you've trained yourself to stay sharp, to stay consistent, to stay disciplined, you reduce the chances of failure.

Coming into the shop distracted or unprepared—wrong mindset, wrong energy— is dangerous. Especially when you're working around heavy equipment. You can't half focus. You can't be thinking about your phone or your hangover or whatever else is going on.

I run a pretty tight ship when it comes to safety. Probably to the point where some of the guys think I'm nuts, but I make them all watch those gnarly industrial-accident videos—guys getting caught in machines, hands crushed in presses, dudes spun up in lathes. It's graphic, but I think it's necessary. If you're gonna be around dangerous tools and machines, you need to respect them. You should be a little scared.

I had to fire a kid maybe six months ago as I write this—for showing up hung over on a Tuesday. I couldn't let him be around the machinery like that. Like, if you wanna party, that's your business. But don't bring that into my business. You're not just putting yourself at risk—you're putting all of us at risk. And I've had a few like that—guys who sneak off to drink during the day, thinking no one notices. I notice.

Thankfully, I haven't had any major accidents in the shop—at least not from the guys. But I've hurt myself by being overtired and stubborn.

There was a time when I was working on a Funny Car team, doing twenty-six races a year, and still running my shop during the week. Monday to Thursday, I'd be in the shop. Thursday afternoon, I'd fly out for a race. Do that for five years straight, and it wears on you.

One Monday, I'd just flown back in, got home at like 1 am. Dead tired. I told myself: *Don't work today. You're wrecked. Take the day off.* But I had a fender to make for a customer, and I didn't listen to myself. I got up early and went into the shop anyway.

I was sanding the center weld on the fender, turned around, walked past the belt grinder—and just barely brushed it. Ten steps later, I realized something felt wrong. I looked down, felt my glove . . . and I could feel the tip of my finger detached inside it.

Just like that. Gone.

I put the finger on ice, drove myself to the ER, ended up needing hand surgery. They saved the nail bed and reattached the tip—it's about three-quarters of an inch now. And yeah, I was back at work the next day. Welding. I don't even know if that was smart or just me trying to prove I was a tough guy. Probably both.

But the point is—it didn't happen out of nowhere. I was tired. I knew I was tired. I told myself over and over to take the day off. And I didn't. That's when this stuff happens: when you're not at 100 percent.

When you do this kind of work—precision work—you have to be sharp. Focused. Rested. Fed. Hydrated. I've gone hours without realizing I had to piss, just so deep in the zone. And yeah, it sounds funny, but it's not great. You've gotta train your brain: If you need to eat, eat.

If you need to rest, rest. Pay attention to what your body's telling you.

Because this work—on the level I try to do it—is not like playing a video game. You don't just "respawn" if something goes wrong. It's real. The machines don't care if you're tired.

I also watch what I eat. I do intermittent fasting—I don't eat until noon—and I stick to a keto diet: no bread, no sugar, nothing that spikes my blood sugar. Food is fuel for me. When I stick to this plan, I feel sharper, healthier, and more capable of doing my best work. The only thing I struggle with is sleep. Most nights, I'm lucky if I get five solid hours. But even with less-than-ideal sleep, I know that staying disciplined keeps me moving forward. My body may not always get the rest it needs, but my mind stays focused because I've built habits that support it.

I won't lie—there are days when it's hard. Days when the grind seems overwhelming. But that's when you rely on the routine. The routine doesn't care if you're tired or frustrated. It just keeps you moving forward, one step at a time. Even when your body's telling you to stop, the routine keeps you in motion. It creates consistency. And consistency is key.

Precision mindset. Precision hands. Precision work.

WHY ROUTINE MATTERS: ELIMINATING DISTRACTIONS

Why is this routine so important? Because discipline eliminates distractions. When you've got a family, a business, and life's inevitable drama—financial problems, relationships, kids—you can't afford to let those distractions bleed into your work. My job requires 100 percent focus. If my mind wanders for even a split second, I could cut my hand off, or worse. In this kind of work, stress isn't just annoying—it's dangerous. So you have to develop the ability to block it out. It's not easy, but it's necessary.

When you've spent a long time working on something, the work becomes second nature. But you don't take that for granted. You develop an awareness that allows you to stay focused on the task at hand, no matter what else is going on in your life. I've had my fair share of problems, but I've always made a conscious effort to shift my mind away from stress and back to my work.

You have to compartmentalize. Work is salvation. It's the one thing that can bring you back to center when everything else in your life feels out of control.

People think that stopping work—taking a break—will solve their problems. It doesn't. Losing your job or abandoning your purpose just compounds everything that's wrong. Work should be sacred. It's your safe space, your escape, and your salvation. There's nothing wrong with taking time off when you need it—but when things get tough, you don't quit. You double down on your purpose and discipline. You show up, day after day, and keep working. Because if you can stick to your purpose during the grind, you'll come out stronger on the other side.

When everything around you feels chaotic, work can be the one thing that gives you clarity. It's the foundation that helps you weather the storms of life. Whether it's building something from the ground up, repairing something that's broken, or just putting in the hours to create something of value—it brings order to the chaos. And that's why routine matters. Because when life is unpredictable, your discipline is the one thing you can count on to keep moving forward.

EXPANDING YOUR SKILLS AND STAYING DISCIPLINED IN YOUR WORK

One of the most exciting opportunities as a creator or craftsman is figuring out how to take the skills you already possess—skills you've honed over years of practice—and channel those into new products, new ideas, and fresh opportunities. For me, that opened a window in 2002 to culinary products: cookware, knives, cutlery, and heirloom-quality kitchen tools. At the time, I was heavily invested in making motorcycles and guns—industries that come with a certain amount of baggage. Some people see bikes and think about accidents or danger. Some people see guns and instantly associate fear or controversy. Those products always have a niche, but they're not for everyone. That's why I found it refreshing to make something universal—something everyone uses and needs—like cookware.

Everybody eats. That's the common denominator. And some people don't just eat; they value the act of preparing meals with tools that feel special—heirloom-quality pieces like knives or pots and pans that can be passed down for generations. Creating those kinds of products still allows me to put my craftsmanship and precision into something, but the audience expands. It's not about fear, it's about family. It's about legacy. There's something satisfying about seeing someone cook a meal with a knife or pan you made by hand and knowing it's a tool that will outlast them, handed down to their kids or grandkids. That connection to people's lives, to their daily rituals, is something that money can't buy. There's a deeper, more meaningful satisfaction that comes from knowing you've created something that will be a part of someone's life for years, something that has intrinsic value beyond just the dollar amount on a price tag.

And that's the beauty of craft—you can take your skill and expand it into different areas. Once you build the discipline and foundation of your craft, you can branch out. The possibilities are endless. You don't have to be boxed in by what you started with. The world is wide open, and it's up to you to explore it. Every skill you learn builds on the next. It's about expanding your horizons, trying new things, and challenging yourself to push your craft further. But it all starts with discipline—the willingness to put in the work, day after day, to refine your craft and build your expertise.

7

FUEL SYSTEM

The 1957 Cadillac fuel system is very simple. A gallon fuel tank is located between the rear frame rails, below the trunk floor. A 3/8" steel hard line gravity-feeds fuel forward to a mechanical fuel pump on the side of the engine. That fuel pump is a diaphragm type that has a rod actuating it off a lobe on the camshaft. This action creates suction from the fuel tank and feeds it to the carburetor. Starting your own business and creating things with your hands also needs a fuel system. That fuel is money, and that money comes from customers.

This is the most delicate tightrope you will be forced to walk in this entire endeavor—always needing fuel to keep things in motion and moving forward. This means you are going to have to deal with all kinds of personalities and characters. My customers have always run a wide spectrum, from top lawyers running the China wing of Goldman Sachs to Compton dope dealers who pay me with trash bags jammed with five- and ten-dollar bills. I can tell you that getting $100K in fives and tens is a lot of cash. When I took some to the bank to deposit it, the bank manager came out and told me, "This cash smells like weed." I was like, "Yeah, well, it's still cash."

When business was booming in the early 2000s, I was selling a lot of bikes, fenders, and parts. I thought it was a good idea to hire a salesman at the shop to talk to customers when they came into the showroom. So I hired this little smart-ass named Trent. He was a typical mousse-haired sales guy who could easily be working at Best Buy. One Thursday, while I was working up in the shop, I glanced outside and saw a taxi pull into the parking lot. What appeared to be a homeless man got out of the back. He had crazy hair that looked like Sideshow Bob's from *The Simpsons*. His pants were at least five inches too short, and he used a bungee cord for a belt. I thought that was kind of weird to see, but I went back to work. Long Beach at that time had about 4,000 homeless people, so I was used to seeing them all day, every day—I just had never seen one get out of a cab. About thirty minutes later, I noticed the cab was still parked in the parking lot. I walked out to the balcony in the front of the shop that overlooked the showroom. I saw the homeless man standing in the middle of the showroom with his arms folded. He was all by himself, just standing there. So I walked down the stairs and asked him if he needed any help.

He was completely tweaked out; his jaw was grinding side to side. He asked me, "I'm looking to buy a bike. Do you have any for sale?" I told him I only had one bike available. It was a Harley we took in on trade and customized, and we were just finishing it up. He asked if he could see it. I said yes and walked him to the back of the shop.

We walked right by Trent, who was sitting at his desk looking at a magazine.

We looked over the bike, and Duane, as he called himself, asked how much I wanted for it. I told him $75K. He said he'd take it and asked if he could pay in cash. I said yes, sir. He also requested that I put a passenger seat and passenger pegs on it. We went into the showroom store, and he bought $2,000 worth of stuff and asked if we could paint two helmets to match the bike. No problem; we were on it. We went up to my office, and he pulled out a paper bag full of cash and counted out $77K in hundreds. With my ego fully intact, I asked him how he heard about me and my shop. He said he never heard of me or the shop. His cab picked him up at the Hollywood Burbank airport, and he got off at the wrong exit. He saw the bikes in the window and told the driver to pull in.

In an effort to find out where he got the cash, I asked Duane where he worked. He told me he didn't work and was homeless in Vegas. He was panhandling on the Strip yesterday, and someone gave him $100. He took it straight into the Venetian casino and put it down on the roulette table. He hit a triple zero and won $400K! He said he flew to Long Beach to see his sister and, obviously, made a pit stop to get some crank. He told me he was headed back to Las Vegas because the Venetian comped his room. Yeah, no shit, they want their money back. Duane asked me if he could come back Saturday to pick up the bike; I told him we would have it done. He got in the cab and left.

He showed up Saturday afternoon, this time in a black Lincoln Town Car with a Black driver who was holding his bag of money and had a very suspect smirk on his face. He also had a lady friend and her two kids with him. She was obviously a hooker—one not very good at her job. She was missing teeth and had really greasy hair. All four of them were wearing brand-new Harley clothes; they had stopped at the Harley dealer in Victorville on the way in from Vegas. The kids were running around the shop like monsters in their brand-new Harley boots and leather jackets, touching everything. Duane was still tweaked out, along with his new lady. Both were grinding their teeth and scratching holes in their arms.

In the showroom, I had a custom '49 Chevy Fleetline that was slammed with wide whitewalls. Duane saw it and said he had always wanted one like that. He asked how much. I told him $40K. He said he'd take it. Then he went into the showroom store again and bought another $2,000 worth of West Coast Choppers clothes. We went up to my office, and he had his driver (who was still holding his money) count out $42K in cash. Duane asked if he could come back the following Thursday to pick up the Chevy. I said no problem. He then put on his freshly painted helmet and got on the custom Harley and rode off. The girl and her kids got in the Town Car and followed him. All headed back to Vegas.

Tuesday morning, I woke up to fifty missed calls from a 702 Vegas number. They left no voicemails. I figured if it was important, they would call back. Later that evening, I got a call from a different 702 number. It was a lady with a scraggly cigarette voice. She said, "Is this Jesse?" I said yes. She told me her name was Dixie, and that last night Duane got arrested in front of the titty bar she works at "because he had warrants." She wanted to know if I had the title for the Harley because they impounded it. I told her I signed it over and gave it to Duane when he picked it up. I watched him put it in his pocket. She said okay and hung up.

Thursday rolled around, and Duane never showed up to pick up the '49 Fleetline. Late Thursday evening, I got a call from another 702 number. This time it was Duane. He said, "Man, that bitch stole my last $50K and split, and they are kicking me out of the hotel. Can you Western Union me some money, and you can just keep that Chevy?" I told him I could only send $1,200 through Western Union at the liquor store down the street. He said okay. So I walked down and sent him the money. I never heard from Duane again.

For a while, I felt really sorry for Duane. Man,

he lost everything. But when I really thought about it, he was homeless and lived like a king for seven days. He got drugs, a custom West Coast Choppers Harley, a hooker, a Fleetline (almost), and a free suite at the Venetian. Plus all his food and booze. He was living it up to the max. So what's the moral of this story? If that little asshole Trent had not judged a book by its cover and had helped Duane when he came in the shop, he would have made his commission on all the stuff Duane bought. Instead, I kept all the cash.

Customers come in all shapes and sizes. Take really good care of them, and they will always fuel your business.

The adage "Don't judge a book by its cover" comes to mind when thinking about this story. We will not be judged on how much money we make, what our title at our job is, or even how many hours we work in a day. We will be judged on how we treat people. Don't let your pride or sense of self-worth get in the way of how you treat the people you interact with daily. The blue-collar life comes with its own stereotypes—people expect you to be gruff, surly, or grumpy. However, I've found that being kind and fair with each person, whether they are rich or poor, works best. You don't have to be a pushover to do that; you can still hold people accountable, follow policies, and be an expert, all while maintaining a friendly demeanor. Teddy Roosevelt once said, "Walk softly and carry a big stick." Do the same in how you conduct your business.

Just like the 1957 Cadillac's simple yet effective fuel system, which relies on gravity and a mechanical pump to deliver fuel to the engine, your business needs a steady stream of income to stay in motion. The story of Duane underscores this point vividly. Despite his appearance and questionable background, Duane turned out to be a lucrative customer who purchased a custom Harley and a classic Chevy, paying in cash and spending generously.

And that's the importance of treating every customer with respect and not judging them by their appearance. Had Trent, the shop's salesman, not dismissed Duane based on his look, he would have earned a significant commission. Had I dismissed him, my business would have lost a significant amount of fuel. The key takeaway is that customers come in all forms—be they high-profile professionals or down-and-out individuals. Taking good care of every customer, no matter their background, ensures that they continue to fuel your business.

ALWAYS GIVE MORE THAN EXPECTED

Cutting corners is one of the biggest mistakes people make. They underbid a job, realize they can't make much money, and then dumb down the work to match the price. I've never done that. I've never once done something half-assed just because of what I was getting paid. If I take on a job, I'm going to do it my way, the best way I know how. This mindset has served me well over the years.

It's easy to get caught up in trying to make a quick buck, especially when you're just starting out. You think, *I'll cut a few corners, make the deadline, and get paid faster.* But what does that do? It damages your reputation. If you're not giving your best work, word will get around. It's a small world, and people remember the jobs that fall short. The reputation you build is worth more than any quick paycheck.

This is one of the core values that has driven my success. When you commit to giving more than expected, you don't just build a product or service—you build trust. And trust is everything. Trust is what keeps customers coming back. They know you won't take shortcuts, that you'll put in the work no matter what. In an age where so many businesses are looking to cut costs and maximize profits, the ones that stand out are the ones that go the extra mile. The ones that give their customers more than they expect.

And here's the kicker: When you give more than expected, it doesn't just build customer loyalty—it builds employee loyalty too. If you're leading a team, they'll notice whether or not you're truly invested in the quality of your work. If you're not cutting corners in your own work, they'll follow your example. Leadership isn't about barking orders from an office; it's about showing up every day and setting the standard. If you want your people to work hard, they need to see you doing the same. That's how you build a culture of excellence. You can't cut corners and expect your team to give it their all. You have to lead by example.

One of the key ways to build that kind of loyalty is to truly care about what you do. Customers and employees alike can tell when you're phoning it in. But when they see you putting your heart into every job, they know they're part of something special. It's about more than just doing a good job—it's about delivering a result that exceeds expectations, every time. And when you do that consistently, you build a legacy.

I think this was before I ever did television— late '90s, maybe. I had just moved into a new shop in downtown Long Beach. It was 10,000 square feet, which felt massive compared to the 1,500-square-foot space I was in before. I remember walking in and hearing my footsteps echo and thinking, *What the hell are we gonna do with all this room?*

Pretty soon we filled it—fender production, frame production, selling to distributors. Materials and equipment everywhere. Shipping pallets going out the door. I honestly couldn't even tell you how much money I was making at the time, but I was in all the magazines, and the momentum was real.

One day, I got a call from my godfather, Barry Weiss—yeah, the dude from *Storage Wars*. He and my dad were business partners back in the '60s and '70s. He saw one of the magazine features and said, "Damn, looks like you're doing great. Have you been sued yet?" I was like, "Wait, what?" And he goes, "That's how you know you've made it."

And honestly? He's not wrong. No one's

suing a guy with nothing. You only get targeted once there's something worth taking.

That kind of thing still happens. I call it the "Jesse Tax." Anytime I buy something, hire a contractor, whatever—I have to vet people hard. Because some folks see the bikes or the guns I build and just assume I'm made of money. Like, "Oh, he builds a $250,000 gun? Cool, let's tack an extra $10,000 on this invoice." They think they're slick. But I see it.

Meanwhile, I treat my own customers in the complete opposite way. I've got clients who are well-off—guys like Manuel Hidalgo, who I'm building a '61 Olds for right now. I built his first bike back in '96 or '97. Nearly thirty years of coming back. Why? Because I always go above and beyond. I always try to give people more than what they pay for.

That mindset is rare. Especially in the luxury space. Too many people are thinking, *How do we screw this guy and take his money?* Never thinking long-term. Never thinking, *Will this guy come back?* But here's the truth: People with money got there for a reason. They've got business sense. You burn them, they'll never come back.

So I do business with honesty. Always have.

I think that comes from watching my grandma. She was an antique dealer back in the '50s, before my dad. Selling rare glassware—Steuben, Tiffany, RS Prussia, all that—at the Rose Bowl swap meet. She had beautiful stuff, but she always priced it fairly. And she always sold everything.

My dad, on the other hand, ran more on charm and white lies. Not a bad guy, just had a different way of doing business. Maybe part of me wanted to be the opposite of that. I've always felt like if you make a great product and stand behind it, you don't have to gouge people. You don't have to play games.

I still wrestle with pricing. Sometimes I underprice because I'm worried about perception—I don't want anyone to feel like I'm trying to take advantage. Yeah, I need to cover my costs and make a profit, but I don't want to be part of that Jesse Tax people assume is baked in to everything I do.

That brings me to the real kicker: the keychains.

The biggest headache in my business? Not the guys spending $250,000 on a gun. It's the guys spending twenty-five dollars on a keychain. I could build a one-off Damascus pistol for a quarter mil, and I'll never hear a word. The client trusts me, lets me do my thing, and is thrilled when it's done.

But the twenty-five-dollar keychain? That's where I get the emails:

"Where's my stuff?" "You stole my money!" "You're a scammer!"

And I get it—it's not really about the keychain. It's about investment. Someone who drops $250K has a deep level of trust. They know what they're getting, and they're in it for the craftsmanship. A twenty-five-dollar keychain buyer doesn't have that kind of trust built up, so they're quick to assume the worst.

Hell, sometimes they buy ten or twelve just to flip them for profit. I created these things to give people a chance to own something handmade by me at a reasonable price. But for some, twenty-five dollars isn't meaningful—it's no different than some cheap plastic gadget off Amazon.

And that's fine. I'm grateful for every sale, for every fan, for every bit of support. But it just goes to show you—it's not always the big-ticket clients who cause stress. Sometimes it's the smallest transaction that makes the most noise.

THE LONG-TERM GAME: BUILDING RELATIONSHIPS

This kind of dedication, of course, takes time. But in a world that's constantly changing, the people who last are the ones who are focused on the long-term. Business isn't about short-term profits or quick wins. It's about building lasting relationships—with customers, with employees, and with your community. The real value comes when you've earned the trust of the people you serve, and they trust you enough to come back year after year.

Building those kinds of relationships takes time and effort. You need to show up, day in and day out, and demonstrate that you're in it for the long haul. Trust doesn't happen overnight—it's earned through consistent, high-quality work and a commitment to doing the right thing. Whether it's your first job or your thousandth, if you put in the effort to deliver quality, people will remember it. They'll want to come back, and they'll tell their friends. Word of mouth is one of the most powerful forms of advertising there is.

You're not just building a business; you're building a community. You're creating something that people believe in and are proud to be part of. That's when you know you're doing it right. When you reach that point, it's not about the money anymore. It's about the relationships you've built, the legacy you've created, and the impact you've made on the people around you.

8

OIL AND LUBE

The 1957 Cadillac has twenty-eight lubrication points. Two of these are short-term service points for engine oil and automatic transmission fluid. These will need to be checked and changed at regular intervals. The other twenty-six service points are for long-term maintenance, and mostly consist of zerk fittings for grease on suspension joints that have limited movement compared to engine bearings that spin thousands of times. Consistent scheduled routine maintenance will keep things functioning properly for thousands of miles.

Deciding to take the leap into a career of working with your hands makes you and your body the engine in your life, just like that big dual-carb 365-cubic-inch engine in the Cadillac. The better you take care of it, the better it will perform and the longer it will last. Taking the dive into something challenging and creative is very intoxicating. Once you start using your hands to build something out of nothing, you won't want to stop. This is a good thing, because working with excitement and enthusiasm will lead to high-quality work and a positive work ethic. However, be warned: Working relentlessly without balance will backfire when things go wrong. Looking back at my thirty-five years in business, whenever something bad happened, it was usually because I was tired. When you work tired, frustration sets in, and mistakes happen.

In the late 1990s, I was building a Panhead for Bob Kay, the president of one of our biggest parts distributors, Tucker Rocky in Dallas. Bob was a great guy, planning to display his new bike at a major motorcycle dealer show in Cincinnati, Ohio. With a tight build window, I was under pressure to finish and ship the bike on time. Late one night, while working alone on the final assembly, I felt a 1/4-20 screw strip in the threaded hole on the head. Instead of stepping back to figure out a solution, I tried to muscle it out with a bigger screwdriver. The screwdriver slipped, leaving a deep scratch on the stainless

Panhead cover. Frustrated, I punched the side of the gas tank, causing the bike to rock and eventually fall off the lift. The fall dented the tank and broke the hand controls. Rather than taking a moment to calm down and address the problem, I created several more issues to fix, including pulling the motor to repair the Panhead cover and replacing the broken hand controls. I had to make a special tool to tap out the dent without damaging the paint. Although the paint was intact, the situation was far from ideal. Ultimately, the fault was mine for not recognizing my own fatigue and frustration.

This act of self-awareness and discipline extends beyond being tired and overworked; it applies to your mental and physical health. If you are not 100 percent into what you are doing, it simply won't work. Distractions can affect your performance. Hard work and long hours must be complemented with proper rest and nutrition. As mentioned in chapter 4, Gene Winfield's diet and work ethic highlight this principle. Eating a heavy meal for lunch will inevitably lead to a sluggish afternoon. I didn't fully appreciate how much my health and physical strength contributed to my work until I began training some of the guys at my shop.

One kid, Luke, struggled with basic sheet metal work because he wasn't strong enough to handle the equipment, and his smoking didn't help. I realized that heavy-hammer-and-forge work requires strength and good health. I now make a concerted effort to eat healthily, avoid sugar, and go to the gym five days a week, ensuring I don't overtrain. A gym injury can also hinder your work. The bottom line is that you need to keep your engine healthy.

Just as the 1957 Cadillac relies on regular lubrication to keep running smoothly, your body needs proper care to perform at its best. Diving into a hands-on career is exhilarating, but working tirelessly without balance can lead to mistakes and frustration, the way my fatigue while assembling that Panhead led to a series of errors, including a dented tank and broken controls. This fiasco could have been avoided by taking a step back and addressing the problem calmly. Self-awareness and discipline are crucial—not just for managing work tasks but for maintaining your physical and mental health. Proper rest, nutrition, and exercise are essential to keeping your "engine" in top shape.

9

MILEAGE

Success is just a lesson in how far you can go on the precipice of failure. Most of the really successful people I know always cry about how broke they are. I've always been a little suspect of customers complaining about how expensive fuel is for their private jet. However, as more time has passed in business for me, I've come to realize how true it can be to be successful and broke. In the early 2000s, West Coast Choppers was going full throttle on all cylinders. We had over 200 employees, a successful global parts business, we were cranking out two to four custom bikes a month, and we were doing $100K a week in online merch sales—and this was before my deal with Walmart started. At the end of the day, my monthly overhead was $405K. That's payroll, rent, power, lease payments, and insurance.

Having started out in my garage a decade earlier with zero overhead, $405K was just insane to me. This is paralyzing mentally and financially. I was always in such a scared state of not being able to make the monthly nut, it was very difficult to move forward with new projects and products. I, in a sense, became a mid-level manager. I was more worried about who was stealing toilet paper and why we were spending $100K a year on packing foam than about my skills and craftsmanship. I really think this kind of fast success is the death knell for any business if you don't get it under control. Amazing things had to get bigger and more expensive for me to step in and make a change.

In 2009, I took a trip up the California coast with my son Jesse Jr. We rode a brand-new Harley Ultra Classic "Shriner edition" with a sidecar. We took the 101 all the way up the coast and stopped to camp overnight in the redwoods. When we made it up to the Central Valley, I noticed the bike was handling weird on the freeway. The steering was starting to pull to the left. I pulled into a gas station and noticed the sidecar frame was cracked and the whole thing was bending in the middle. Since we were in the Central Valley, I called my old friend Cole Foster in Salinas. I asked if I could limp the bike over to his place to fix it. He said no problem, and Jesse Jr. and I headed that way. Cole builds custom cars and bikes in his home shop. His house is a cute little two-bedroom California bungalow-style home with a huge 3,000-square-foot shop behind it.

We got the bike in the shop and started to assess the damage on the sidecar frame. It was pretty bad. The whole frame casting was broken, and the sidecar body was lying against the crash bars on the bike. The whole thing needed to come apart, be jacked up, and welded. We looked at a couple of ways to temporarily quick fix it, but I really didn't want to take chances with the safety of my son in the sidecar. If it broke all the way off, that would be bad. So we decided to catch an Amtrak train back down the coast to Los Angeles the next morning. Cole took us to dinner and let us stay in the loft in his shop. Catching the Amtrak at the Salinas station the next morning, part two of this adventure was just beginning.

Jesse Jr. and I were cruising along in the dining car when we felt a huge jolt. Then the brakes immediately came on hard, and the train came to an abrupt stop. Looking out the window, I could see the ground was littered with tomatoes. Half-jokingly, I told Jesse we hit a tomato truck. It was true; the train had driven straight through a tomato truck stuck on the tracks. They let us all off the train, and we were able to see the damage and take some pictures. We had to wait two hours for another train to come pick us up and head down to LA. I told Jesse Jr. this father-son trip was a total train wreck.

Despite all the drama on our trip, I could not stop thinking about Cole's shop behind his house. It was perfect. To be able to walk outside and go right to work, with low overhead, was ideal. I thought about it and thought about it until a year later, when I finally picked up and moved my shop to Austin, Texas. I went from six buildings to one, and from 200 employees to twelve. I eliminated everything in my life that was a distraction from working. I successfully lowered my overhead to a sustainable level. By sustainable, I mean if the chips are down and I need to cover payroll, I can hustle and make it happen in a couple of days. When the overhead is half a million dollars a month, the hustle is very different, and it requires shady loan sharks and creates even more debt.

So when I say success is how far you can stay

on the precipice of failure, it means that to make things work for a really long time, you have to be willing to sacrifice all of your resources to the financial health of your business.

The most important thing for me is to make sure everyone who works for me gets a check every two weeks. I consider success to be knowing I have the money in my account to cover bills and payroll. Nonsuccess is when I have to scramble and wheel and deal to cover things. I've cut it so close that money was being deposited in my account even as my employees were depositing their payroll checks. Thank God for overdraft protection. This kind of revolving cash register business is very stressful. Most people can't deal with it, and they shut down. I have always dealt with it with a certain amount of confidence that comes from experience. If you work hard and make a good product, things will always magically work out. A customer will stop by with a check (or, in my case sometimes, a bag full of cash), orders will come in. As long as you keep working, things will always be okay. This is also why keeping overhead low and sustainable is important. Growing things to a huge size and blowing up costs just for the perception of success is not a good path and will usually end itself. Be lean and mean with fewer moving pieces. Set yourself up to handle anything that comes your way.

Success often teeters on the brink of failure, and my experience proves this. Back in the early 2000s, despite West Coast Choppers thriving with over 200 employees and hefty sales, that crushing overhead of $405K a month made me constantly anxious. I was so bogged down with managing costs and worrying about trivial issues that I lost focus on my craftsmanship and the excitement of building. That tomato train-wreck trip with my son highlighted the value of simplicity when our Harley broke down and I saw Cole Foster's streamlined operation. The trip led me to reevaluate everything, eventually moving my shop to Austin with a leaner setup. Now, with reduced overhead and a manageable team, I focus on ensuring everyone gets paid and maintaining a healthy balance. Success is about keeping your business lean and sustainable, handling challenges with confidence, and staying focused on what truly matters.

10

FOUR-DOOR FLEETWOOD

The 1957 Cadillac Fleetwood four-door hardtop was the Cadillac of Cadillacs. Two plush bench seats, with enough room for six passengers. This is the car you wanted when you needed to take everyone for a ride.

The thing is, taking everyone with you isn't necessarily a good idea. I'm going to tell you this up front: Do not hire friends or family in your business. As much as you want to take everyone on this magic carpet ride, this will always end badly. It took me twenty-plus years to figure out why. Making the decision to break away from societal norms and become a craftsman is a very rare and attractive thing. Once you get really good at something and that open-for-business sign flickers on, the hangers-on will flock to you like moths to a porch light.

For me, at first, this was really exciting. My Hackett Avenue garage was a magnet for the neighborhood. I was such a local phenomenon that my first business cards didn't even have an area code on them. I think my overall reach goals were about an eight-block radius. Every day after work and every weekend, my garage had a steady stream of friends and neighbors stopping by to get this fixed, or get something welded, or just for me to show them what I was doing. The best was all the neighbor kids having me modify their bicycles.

One of the guys who used to stop by the shop all the time became a close friend. We rode together for years. Eventually, he came to work for me, and for a long time, I trusted him with everything. Then one day, a customer mentioned that this friend had made a big, all-cash purchase that seemed out of step with what I knew about his finances. Around the same time, I started noticing other red flags—lavish spending, inconsistent paperwork, and cash that didn't seem to add up. Looking closer, it became clear that money was leaking out of the business in ways I hadn't realized. The hardest part wasn't the financial hit—it was realizing that someone I considered a close friend had been taking advantage of my trust. That experience changed how I ran the shop and who I let handle cash from that day forward.

Pretty funny, I think. It's so confusing I don't even know how much money I was out at that point. He was eventually caught in the act and fired. But being that this dude was such a close and trusted friend, I was really hurt for a couple of years. If he had asked for any of this stuff, I would have just given it to him. I was just so confused and betrayed. It took me several years to realize why he did this.

Robert Greene's book *The 48 Laws of Power* gave me the answer I needed. There is a whole chapter about envy. Greene says that if you have any friends in your life, and they are not on the same economic or success level as you, get rid of them immediately. They are seething with envy and most likely hate you. This envy can manifest itself in many ways, including constant praise. That's what this dude did—constant praise. Always telling me how happy he was for me and how great it was that my TV show was successful. In reality, he was stealing behind my back and justifying it because of his envy. He believed he was owed all that stuff and money because I had reached a level of success and he hadn't. Even though it was my wish to take my old friend, who had been there from the beginning, on the magic carpet ride that I was on, it for sure backfired. Lesson learned.

Family can be just as bad—or even worse. My first wife, Karla, hated the idea of me having a shop. She and I got together when I was doing security for bands, and that's how she liked me. In 1993, when I decided to quit everything and work with my hands building motorcycles, she was not on board. The shop was always a source of conflict until our marriage was over.

All wives and girlfriends will say they love your shop and love that you are dedicated to your craft. In reality, most of them really don't. It's very rare to find someone who will fully understand your commitment and dedication to making things work. Most of society looks at shops as dirty, dank places that nobody should want to hang out in. I think some see it as a novelty that wears off pretty fast. I mentioned in a previous chapter that you really have to do all of this for *you*. It has to be your calling in a selfish way. You cannot expect your significant other to jump on board with this. I have come to the realization that I have to be okay with them not liking it at all. Trying to force someone to mirror your enthusiasm about welding and grinding in a dirty shop is pretty unreasonable. I mean, if we were insurance salesmen working in cubicles in a big anonymous building, we would totally understand our partner thinking it sucks. So we have to be okay if the enthusiasm for whatever we're doing isn't mirrored.

I'm not saying all family and friends are bad and ill willed. I'm sure plenty of people have great experiences hiring them. My only real advice is to just be cautious and suspect of everyone. People will put themselves in your path just so they can screw you while you aren't looking.

Despite the shiny exterior of success and camaraderie, behind the scenes, envy and discontent can corrode even the strongest bonds. It's a tough pill to swallow, but the truth is, some folks just can't handle the ride when the road gets bumpy. The best course of action is to keep a wary eye and protect your hard-earned gains. Remember, while you might want everyone to share in your success, not everyone can handle it. Keep your circle tight, stay focused, and always be ready to handle what comes your way.

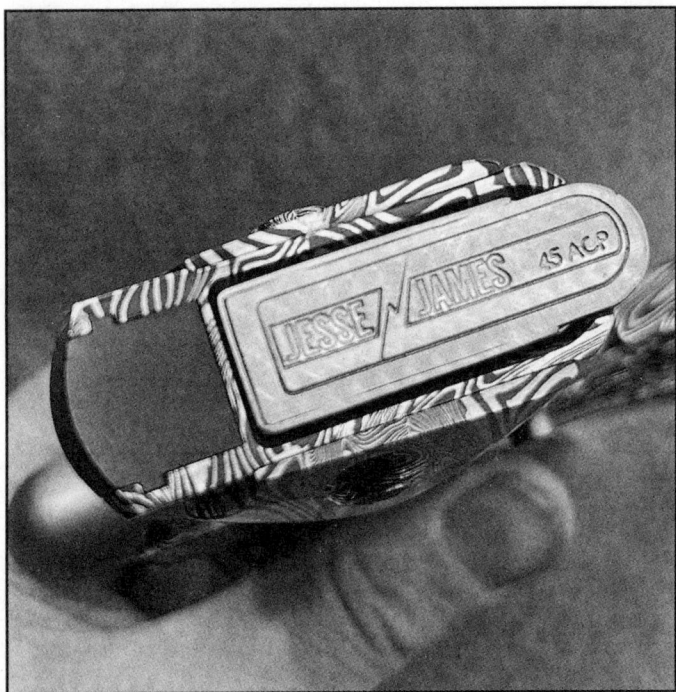

11

AUTHENTICITY: WHY WEST COAST CHOPPERS WORKS

I've seen so many businesses fail because they were built on nothing but marketing and branding. Take Mossimo as an example—it was a clothing brand that looked cool for a while, but there was nothing behind it. It was just a name. There was no guy behind it, no heart, no soul. That's why it disappeared. People saw through it because there was no real passion behind the product.

West Coast Choppers is different. The logo is cool, the bikes are cool, but there's also me. There's a real person, Jesse James, who builds the bikes, touches the work, and makes sure everything is perfect before it leaves the shop. I don't just slap my name on things and call it a day. My hands are on everything—clothing, motorcycles, cars, guns, knives, cookware, whatever it is. I'm involved. Nothing leaves the shop without my approval. That's the engine that drives every one of my businesses: me. And the people who work with me—they know I'm right there with them, making sure everything is done the right way.

In today's world, that kind of authenticity is rare. I look at contemporary artists like Jeff Koons—one of the most successful artists in the world—and I see how hollow it feels. He doesn't make his own work. He designs an idea, ships it off to a foundry, and a team of 100 people welds, sands, and paints the sculptures. Then he comes back, signs his name on it, and takes all the credit. I never want to be that guy. I don't want someone looking at my work and thinking, *Jesse didn't even touch this.* If I ever get to that point, I'll sell everything and walk away. I'm in this because I love the craft, not because I want my name on things I didn't make. When you start outsourcing the soul of your work, you lose what made it special in the first place. That's when it becomes just another product, another commodity in a market full of them.

That's why West Coast Choppers works—it's real. You can feel it in the work. You can see it in the details. When someone buys one of our bikes or wears one of our shirts, they know they're getting more than just a product—they're getting a piece of me. That authenticity is what keeps people coming back. It's what makes all the difference. And as long as I'm still here, still working, I'll make sure it stays that way. Because at the end of the day, people can tell if something's fake, even if they can't always put their finger on why. It's that feeling of being connected to the creator, to the work itself. And that's what keeps customers loyal. That's the foundation that makes a business not just survive but thrive.

PART 3

TRUE CRAFTSMANSHIP

12

RESALE VALUE

The rise and evolution of my shop, West Coast Choppers, has been an insane, unpredictable roller-coaster ride. I have evolved, changed, and gained perspective on my life as a craftsman. I in no way think that I am qualified to teach some kind of college business course, but I do hope that by sharing my three and a half decades of experience, you can find your path. Honestly, I think some of the stories I share are more of a deterrent than anything else. I'm comfortable with that, and I hope you can gain insight from my mistakes.

The choice I made in 1993 has been the best thing I have ever done in my life. Things were pretty bleak after my college football career ended. I can remember praying to God to let me do something or let me invent something. I just needed a purpose. Little did I know that the thing I needed, I was already doing: working and learning with my hands and building things in any free time. Society wasn't ready to consider it a great career choice, but I was.

When I was putting myself through college and playing football, it was the absolute poorest time in my life. I can remember putting seventy-seven cents' worth of gas in my VW bus to make it to class. There was also a period of time when I was homeless and living out of that bus. Then something happened. I started working, doing physical labor. I attacked my work, and building bikes, with the same aggressiveness and work ethic I had in football. I worked late, long hours. I chose work instead of hanging out with my friends. I put my whole heart and soul into it. Then, just like that, people started giving me money to do stuff. So I did anything and everything to keep things going. Any jobs that came my way, I did. Yes, there have been times of thick and thin, and adversity to overcome. Some of my ventures have thrived and others have struggled. But from the day I moved into that shop on Minnesota Ave, I have never been able to catch up on work and back orders. That fact alone gives me the highest self-value, knowing that I make stuff that has been in high demand for so long.

I think the notion of considering myself a craftsman is relatively new for me. I think I've mostly focused on the work it took to build bikes and products. I never really considered what I was actually doing on a personal level. It was the actual work, and it's always been the work. It's easy to overlook work, skills, and craftsmanship when you have this big, beautiful, candy-painted-and-chrome monster standing in front of you. I never really considered the fact that I needed to be making things, and they just happened to be custom bikes. Ten years ago, I went down the path of building guns. I get the same fulfillment as I do building bikes. It's just a different rabbit hole for me. Trying to build a better mousetrap and make it look like it was built with craftsmen's hands. So, hopefully, reading this, you will realize that you are where you need to be. Those things at the end of your arms—your hands—need to be doing something. The easy way out is to just get a job in an office, make a dependable paycheck, and be miserable. That's what everyone does. That's never been for me. I read a quote a long time ago saying that using your hands to feed yourself and your family is the most noble and honorable way of life. I agree with that 100 percent.

On this journey through the highs and lows of running a custom shop, the key takeaway is the profound impact of dedication and authenticity in craft. From the roller-coaster ride of West Coast Choppers' rise to the personal growth and insight I gained along the way, the lessons shared are both raw and enlightening. The evolution from a garage-based operation to a full-blown enterprise demonstrates that success comes with its own set of challenges, including the temptation to mix business with personal relationships. The stories of betrayal and hardship underscore the importance of maintaining professional boundaries and focusing on the core values of hard work and integrity.

Embracing the craft—whether it's building bikes, guns, or any other handmade work—requires more than just skill; it demands passion and perseverance. The book reflects on the profound satisfaction that comes from creating with one's hands and staying true to one's calling. It's about finding purpose in the work, even when the path is fraught with obstacles. The final message is clear: True success is not just in achieving financial milestones but in continually honoring the craft and keeping the spirit of craftsmanship alive.

13

REAL LIFE VS. YOUTUBE: THE PHILOSOPHY OF TRUE CRAFTSMANSHIP

The divide between real, hands-on craftsmanship and curated, online personas has never been wider. We live in an era when influencers can create the illusion of expertise, but as anyone who's set foot in a real workshop knows, you can't fake skill when it matters. True craftsmanship demands more than surface-level appearances—it requires substance, sweat, and a relentless pursuit of mastery. This isn't just about blacksmithing, welding, or mechanical work; it's about a philosophy. It's about the mindset that separates real-world grit from internet gloss, the difference between those who build with their hands and those who pose for clicks.

It's easy to look at polished YouTube videos of experts performing intricate tasks and think, *That's the way it's done.* But there's a gap between what's shown and what's truly required to master these crafts. It's the difference between watching someone pour molten metal into a mold and actually *doing* it, feeling the heat of the forge, understanding the tools that shape the final result. This divide is where the real challenge lies—the bridge between watching and doing, between pretending and being, between virtual and tangible. We're not just talking about metalwork or woodworking; this is about the whole mentality that goes into doing something well, that goes into becoming good at what you do.

YOUTUBE VS. REALITY: A STORY OF SUBSTANCE

Let me tell you a story—a real story—that perfectly illustrates this point. There's a guy, a blacksmith from Sweden, whose work I've followed for years. He makes cool, neat stuff in his little house out in the Swedish countryside, and he's carved out a solid online presence. On YouTube, he looks confident, skilled, and dedicated to the craft. But one day, he walked into real life.

This guy visited my friend Roger Lund's shop in Skog, Sweden—a place steeped in history and tradition. This isn't your "cute blacksmithing" shop with polished aprons and aesthetic photo shoots. Skog is where the real stuff happens: iron ore forged into industrial tools, massive components for ships, Volvo prototypes, things you wouldn't believe were still done by hand. Roger's the man behind it all. At fifty-seven, he's a machine. The guy's ripped, relentless, and can swing hammers that most people couldn't even lift. Roger doesn't dabble—he owns his craft.

When the Swedish blacksmith stepped up to work one of Roger's hammers, reality hit hard. I've been there; I've seen it. These hammers aren't for show—they're tools that demand respect and skill. One wrong move, and you're wrecking steel or worse: wrecking yourself. The blacksmith gave it a shot, and Roger stood in the background, watching. It didn't take long before Roger shut things down mid-attempt, and let the guy step back. No words needed to be spoken. The look on Roger's face said it all—disappointment, frustration, disgust.

That's the difference between YouTube and real life. Online, you can edit mistakes. You can curate confidence, crop out failure, and manufacture an image. But when you're standing in front of a 1,000-pound hammer or a 2,000-degree forge, the truth comes out. You either have it or you don't. What you see on YouTube doesn't show the years of hard work, the repetitive failures, or the learning curve that's required to become truly skilled. It's a façade that misrepresents the harsh reality of craftsmanship.

A CULTURE WITHOUT SUBSTANCE

Influencers today are all about appearances. They've mastered the art of creating just enough skill, just enough polish, and just enough editing to look like they know what they're doing. Their work rides trends and algorithms—clever cuts, viral challenges, and dances with tools in their hands—but it's all a show. Scratch beneath the surface, and there's nothing there.

This emptiness isn't just in the workshop—it's creeping into every creative industry. Publishers throw book deals at influencers who've never actually built anything meaningful. Their work gets packaged and sold as if it's something real, but it's just content—fluff designed to make a buck. That's not who I want to be, and it's not the kind of work I respect. These online personas might look impressive, but they miss the essence of what it means to be a true craftsman: the journey, the sweat, the constant learning, and the quiet confidence that comes from real experience.

Behind every true craftsman is a story of mistakes made, lessons learned, and time spent honing a craft. You can't rush that process. You can't fake it. And it certainly can't be captured in a few slick edits for a social media post.

14

A REAL VISION: CRAFTSMANSHIP FOR EVERYONE

Here's my philosophy for this book: It's about connecting real skills to real people. I've spent my life learning from the best craftsmen around the world. I've been in workshops in Sweden, Israel, Europe, the US—studying blacksmithing, welding, fabrication, electrical work, and more. I've spent hours reading, researching, and practicing these crafts, building a toolbox of knowledge and experience that's taken a lifetime to earn.

Most people still see working with your hands as something lower—something beneath them, something blue collar and overlooked.

Society tells the office worker with a dream of learning to weld that it's out of reach. But it's not. These skills are for anyone willing to put in the time, get dirty, and build something real.

There's an old myth that craftsmanship is reserved for a special few. But the truth is, these skills are not beyond you. They are accessible to anyone with the passion and commitment to learn. You don't need a special background, and you don't need an elite education. You just need the desire to try, to fail, and to keep going. It's about reclaiming the dignity of work, the satisfaction of creating something with your hands, and the joy of seeing a project come to life.

I remember this one letter I got after *Monster Garage* aired. It was from a fifteen-year-old girl whose dad was a diesel mechanic. She told me that watching the show helped her dad feel respected. That hit me. Mechanics, blacksmiths, welders—these are the people who build the world we live in, and too often, they're dismissed or underestimated. That's the heart of this book. I want to inspire people to see the value in craftsmanship again. I want to bridge the gap between those who dream of building and those who already do.

THE POWER OF GRIT

True craftsmanship isn't about appearances—it's about showing up every day and putting in the work. It's about hours in the shop with no cameras, no likes, and no applause. It's about learning from mistakes, facing setbacks, and refusing to quit until you've mastered something. Real craftsmen understand this, but it's something the internet often forgets. You can't Photoshop skill. You can't cut out the failures and skip to the success.

When you step into a real workshop—whether it's to weld your first bead, forge your first piece of steel, or carve a piece of wood—there's no hiding. It's just you, the tools, and your ability to figure it out. That's where confidence comes from. That's where real substance lives. The internet often tells us that success should be quick and easy, but true craftsmanship teaches us that the road to mastery is long and hard. And that's exactly how it should be.

Real craftsmen are proud of their failures because each one is a step toward improvement. When you fail, you learn, and when you learn, you grow. The only way to really become good at something is to keep showing up, putting in the hours, and refusing to accept anything less than your best. And the best part? That process is something no one can fake.

15

BRINGING BACK THE TRADES: A MISSION

We need to bring this work back into the spotlight. Society needs to value it again. Industrial arts should be in our schools, apprenticeships should be thriving, and people should understand that these are careers that matter. My goal with this book is to fill the gap between the dream of working with your hands and actually doing it. For too long, we've sent the message that manual labor is beneath us, but I'm here to tell you it's the backbone of everything. Without skilled tradespeople, none of the stuff we take for granted—the cars we drive, the buildings we live and work in, the machines that keep our lives running—would exist. These are essential, foundational careers that build the world around us.

There's still a stigma, unfortunately, that these jobs are a last resort. Growing up, my dad would tell me, "Go to college so you don't have to do this." It was always, "Get a degree, get a suit job, and be successful." But what does that even mean anymore? Nowadays, we see kids racking up six-figure college debts to study political science or graphic design, only to realize they can't find work. That's not success. To me, success is someone who opens a bakery in their hometown, serves the community, and bakes for locals. That's real. That contributes to society.

Colleges have become a massive business, and people are waking up to the scam. Look at the beautiful campuses, the tax-exempt property, the thousands of kids paying $40,000 a semester. And for what? You think it's "exclusive" to get into Harvard? Pay the money, and they'll let you in. They don't care. Meanwhile, we're neglecting the trade schools, the shops, the apprenticeships—places where kids could learn real-world skills and enter a workforce that's desperate for talent. The truth is, working with your hands is a path to a rewarding life if you have the right mindset.

LOSING EVERYTHING AND STARTING OVER

For me, football was everything. My plan was simple: Get a scholarship, go pro—end of story. I didn't think further than that because I didn't need to. But when I got in trouble and got arrested, all of it was gone. I didn't have a backup plan. I worked part-time in a shop throughout high school and college, but after everything fell apart, I ended up doing high-security work. I toured with bands, worked for Rick Rubin and Warner Bros., and spent five years burning myself out across Europe.

Even during that time, though, I was an enthusiast at heart. I'd find myself at bike shows, flipping through magazines, visiting workshops. When I was home—which was rarely—I'd be in my garage, ordering parts, fixing bikes. But working 100-hour weeks on security jobs wore me down. I made good money—$100,000 a year—but I aged twenty years in five. The pressure, the burnout, the lack of satisfaction—it all caught up with me. That's when I realized I couldn't keep going on like that. It wasn't a life.

FINDING PURPOSE IN THE GARAGE

In 1992, working with your hands wasn't considered respectable. Every kid in my high school shop class was a stoner trying to coast through an easy program. So when I decided to quit everything and build motorcycles, people thought I was out of my mind. "Why would you do that?" I was making good money doing security, but I hated it. I wanted to work with my hands. I got a job at Performance Machine, working for Perry Sands—a guy who knew my dad. I'd never worked an hourly job before, so when they offered me fifteen dollars an hour, I had to ask my mom if that was any good. "Yeah," she said. So I took it. From there, things snowballed.

I went on to work at Hot Rods by Boyd, one of the best custom-car shops in the world. That job opened my eyes to what was possible. At the same time, I started making motorcycle fenders in my garage, working sixteen-hour days—eight hours at Boyd's shop, eight more at home. Boyd paid me $750 a week, but I was selling $15,000 worth of fenders a week out of my garage.

16

CREATING LEGACY OVER INSTANT GRATIFICATION

The thing people don't realize about the shop life, the entrepreneurial life, or the craft life, is that it's a marathon, not a sprint. Too many people want quick success, instant gratification. They want to launch their business and be rich next week. But the true joy of craftsmanship and business comes from building something that will last, something you can look at and say, "I created that from nothing."

The process of craftsmanship—whether it's welding, carving, or mechanical work—takes time. There's no cutting corners if you want to build something that endures. People often get impatient, wanting results before they've done the work. That's when you see things fall apart. They didn't invest the time, didn't put in the hours, didn't earn the trust and respect that comes with real mastery. And eventually, they'll find themselves frustrated and burned out. Success is something that grows. It takes time, it takes patience, and it takes everything you've got.

That's why you have to love what you're doing, because it's going to get tough. There will be nights when you're just about ready to give up. But it's the love of the craft—the respect for the process—that keeps you moving forward. That's the difference between the people who make it and the ones who never make it past the first roadblock.

If you want to make it in business, in craftsmanship, or in life, remember this: Success doesn't come from shortcuts, it comes from putting in the work, learning from your failures, and never, ever giving up. And when you do that—when you show up, day after day, putting in the hours, the sweat, the heart—people will notice. You'll build something real. Something that matters. Something that lasts.

17

THE VALUE OF SKILL AND MASTERY

Skills are everything. They're your foundation. You can lose your shop, your tools, and your equipment, but no one can take your skills away from you. That's why your focus should always be on mastering your craft. Become so good at what you do that no one else can compete with you. If you're a welder, be the best welder anyone's ever seen. If you're a machinist, do work so precise and so clean that people say, "Only that guy could do this." It's not enough to be good at what you do—you need to be great.

The path to mastery is long, often frustrating, and requires a level of commitment that most people aren't willing to give. It's easy to get distracted by the noise, the shiny new tools, or the latest trend. But the truth is, those distractions won't help you in the long run. Your true strength comes from the depth of your skill. Think about the greatest craftsmen, the legends who have left their mark in history. What sets them apart isn't just that they had good tools or were in the right place at the right time—it's that they were the absolute best at what they did. They weren't cutting corners or chasing after the next big thing; they were focused on their craft, honing it every single day.

A CNC machinist in a massive shop with 300 machines might have incredible technical skills, but if all they're doing is making thousands of the same part over and over again, where's the fulfillment? They don't even know if the parts are good—they just know no one's brought them back. That's not the kind of work I want to do. I want to see the result. I want to know I've created something great with my own two hands. That's why it's so important to have pride in your craft. Don't just do the job—do it better than anyone else. Take satisfaction in the details. If you're making cabinets, make them so good that people say, "Did you see those? No one else can do work like that."

When you do that—when you put in the extra effort, the extra care—you'll never struggle to find customers. People will line up to hire you because they know you're the best. It's like a snowball effect—the better you are, the more business you'll attract. But mastery doesn't come overnight. It takes years of practice, failure, and adaptation. In the beginning, you'll probably make a lot of mistakes. But that's part of the process. Don't get discouraged. Those mistakes are the building blocks of skill. The faster you fail, the quicker you'll learn.

18

BUILDING SOMETHING BIGGER THAN YOURSELF

In the end, it's not just about your shop or your tools. It's about what you leave behind. Are you creating something that will last, something that will continue to impact others long after you're gone? That's the true measure of success. Building a business isn't just about profits or growth— it's about the mark you make on the world.

The people who succeed aren't the ones who chase after money or status. They're the ones who follow their passion, who put in the work, and who care about the details. And when you do that, when you build something from the heart, everything else follows. The customers will come. The employees will be loyal. And you'll create something that lasts—a legacy of skill, craftsmanship, and integrity.

That's the real secret to success. Not cutting corners, not chasing shortcuts, but doing the work with passion, with pride, and with respect for the craft. So when you take that leap, make sure it's something you love. Make sure it's something worth building. Because when you do, it'll be worth every hour, every sacrifice, and every bit of hard work you put into it. And that's what success is all about.

THE VALUE OF PURPOSE AND PUSHING THROUGH THE GRIND

One of the most important things I've learned is this: You can't rely on external validation to feel accomplished. There's no trophy at the end of the day, no pat on the back waiting for you. A lot of the work you do—especially in a skilled trade or a hands-on craft—is fully self-serving. You do it for you. But that doesn't mean it's easy. In fact, the hardest part about monotonous work—hours of repetition, precision, and focus—is finding the payoff in it. You have to know that the grind, the long hours, the bleeding fingers, the strained eyesight, and the exhaustion all mean something. They're not for nothing.

I've worked through so many days when my body felt like it was about to give up. There's a physical toll that comes with putting in the hours, whether it's the calluses on your hands from welding, the aches in your back from hours of standing, or the sheer fatigue from long days that seem to stretch on forever. Trust me, there are days when it gets harder to push through than others. But that's when you need a bigger reason—a bigger goal to keep you going. Your work needs to mean more than just showing up and punching the clock. It needs to fuel your purpose.

If you're stuck in a boring job or a soul-sucking occupation, the last thing you want to do is move sideways into another equally boring gig. That's the definition of insanity: doing the same thing and expecting a different outcome. No, if you're putting in twelve- to sixteen-hour days, you need to have a bigger goal—a real reason to keep pushing. Something that transcends the exhaustion and monotony of the work. Because without that sense of purpose, the grind can easily consume you.

That reason has to be so powerful that you're

willing to endure the grind, even when your body is breaking down. And trust me, your body will break down. At fifty-five years old, I feel it. The long hours I've worked over decades—welding, forging, manufacturing—it all adds up. But I've always been able to push through because I had a purpose. I've always worked out, tried to stay semi-healthy, and kept myself on track with a regimented routine. That routine is the key to surviving the grind.

The trick isn't just pushing through for the sake of pushing through—it's understanding why you're doing it in the first place. Why you're putting your body through that strain. It's the knowledge that each hour, each small sacrifice, brings you closer to something bigger. Maybe it's the satisfaction of building a reputation as someone who gets the job done right. Maybe it's the desire to build something that will last, something your kids or grandkids will be proud of. Whatever that purpose is, it needs to keep you going even when your body and mind are screaming at you to stop. And when you're living with purpose, that's exactly what happens. You keep going.

19

MONETARY SUCCESS VS. FULFILLMENT

For years, I thought success meant money. In California, I had seven buildings, hundreds of employees, and millions of dollars in business. On paper, I was "successful." But in reality, I was miserable. I was a mid-level manager with no connection to the work I loved. My days were filled with meetings, spreadsheets, and making decisions that had nothing to do with my passion. In 2010, I finally said, "Fuck it." I got rid of everything that didn't matter and moved to a place where I could focus on what I love: building, creating, and working with my hands. That move wasn't just about shedding the material things—it was about shedding the idea that those things represented success. I realized that all the money and status couldn't replace the joy I felt when I was in the shop, creating something with my own two hands.

Now, I'm the happiest I've ever been. I wake up every day excited to be in the shop. I'm doing what I love, on my terms. That's what real success looks like—not buildings, not employees, not bank accounts. It's waking up with purpose and knowing you're where you belong. For years, I believed that success meant accumulating wealth, but it took me years to realize that what really matters isn't how much you have but how much you enjoy what you do. No amount of money can replace the satisfaction of building something with your hands, of seeing the finished product and knowing that you gave it your all. It's about connection to your work, your purpose, and yourself. It's about waking up every morning with a sense of fulfillment that doesn't come from what's in your bank account—but from the work you're doing and the legacy you're building.

Money can't buy peace of mind, and it can't buy happiness. I've seen it firsthand. I was making more money than I knew what to do with, yet I felt empty. The key to happiness isn't more stuff—it's about the satisfaction that comes from pouring your energy into something meaningful. You can have all the material wealth in the world, but if you're not connected to your work and your purpose, it's all hollow. That's something you can only learn through experience, through trial and error, through years of chasing the wrong thing. Once you let go of that belief that money equals happiness, you free yourself to pursue what truly matters—what fills your soul, not your bank account.

PLANTING THE SEED OF SUCCESS

This whole thing—starting a business, creating a craft, building something with your own two hands—it's like planting a seed. That seed is you: your skill set, your work ethic, your passion, and your drive to succeed. You plant it in the soil, and every hour you put in, every day you show up and nurture it, that seed grows. And let me tell you, when it starts to sprout—when people start to notice what you're doing and admire the work you've built—it's the most incredible feeling in the world.

The growth of your craft is gradual. At

first, it feels like you're not getting anywhere. You put in hours of work, but the payoff seems small. It's frustrating. It's easy to wonder if all the effort is worth it. But if you stay consistent, if you nurture your business the same way you would a garden, it will bloom. The thing is, you have to trust the process. Trust that the work you're putting in will pay off, even if it doesn't happen overnight.

Most people give up before they see the first signs of growth. They expect immediate results, and when they don't get them, they throw in the towel. That's where the difference lies between those who make it and those who don't. Patience. Consistency. And most importantly, love for the work. You have to enjoy the process of growing your craft, not just the end result.

I don't care what you're making. You could be building wooden coffee cups, crafting leather belts, or baking bread every morning. It doesn't matter what the product is. What matters is that you're doing something unique. You're creating something with value. And when people see that—when they appreciate the quality, the time, and the effort that only your hands can deliver—they'll come back for more. They'll feel the difference between something mass-produced and something made with care, skill, and passion.

The beauty of building something with your hands is that you can see your impact. You can look at the end product and know it's something you made, something that's yours. That's the magic. That's why I pour everything into my craft—because it's a reflection of who I am. I know that every project is a piece of me, my skill, my vision, and my dedication. Whether it's a motorcycle, a knife, or even a custom-built piece of machinery, I can see the fingerprint of my passion in every curve, every detail, every element that I've created.

Let me give you an example: Say you're a baker, and you've got customers who come every single day for a specific loaf of bread. They depend on you. They know that when they walk into your shop, they're getting exactly what they want: the perfect loaf, fresh out of the oven. And one day, your oven breaks. Those customers are going to be upset. They're going to feel like something's missing in their day. Why? Because you've created something that matters. You've built a trust and a connection with them.

That feeling—when people rely on you, when they admire the work you do—is addicting. I'm telling you, it's fuel. It's what keeps me going. When you've poured your soul into your work, it's a special kind of reward to see that people appreciate it. And that admiration? It doesn't just stay with the work itself—it creates loyalty. It creates a relationship with your customers. They don't just want your product; they want your touch, your craftsmanship, your story. That's why it's so important to give everything to every project you take on. Your passion will shine through, and people will come back again and again because they can see the value in what you're creating.

When I finish a project, whether it's a bike, a knife, or a piece of machinery, I'll take it to a show or park it somewhere and step back. I'll stand 100 feet away and just watch. Watching people look at it, lean in, and admire the work—it's like hunting. I feel like a big-game hunter, and their admiration is the prize. I'll hear them talking to their friends about how I made this or how I did that, and that feeling is everything to me. It's not about the ego, it's about the connection. It's about knowing that what I've made has impacted someone's life in a meaningful way.

That's why I care so much about making sure people love what they get from me. I had a guy call me yesterday about a gun I built for him. He wanted all the parts finished in a specific shade of blue, but when he saw it, it had a little bit of a bluish-green tint. Turns out, he went to Duke University, and it had to be Duke blue. So what did I do? I took it back. I'm redoing it. Because I'm not going to let my ego get in the way of his satisfaction. I don't want him to settle. I want him to love it. That's the addiction—it's about that feeling. It's not about the money or the ego or the sales. It's about knowing I gave someone exactly what they wanted—something that only I could make for them—and that they're going to love it. They'll come back to me again and again because they know I care as much as they do.

20

PROTECTING YOURSELF AND YOUR WORK

Success, though, comes with its own challenges. When you build a business, gain a little notoriety, and start making money, you have to be careful. People will step into your path just to take advantage of you—to steal from you, to sue you, to use you. I've been there. I had a shop in downtown Long Beach for years. Then I got on TV, and suddenly millions of people were watching me. And you know what? People came out of the woodwork. I had friends, family, and strangers trying to take advantage of me. It's not something I ever imagined would happen, but it's a reality of success. When you're in the spotlight, people will do whatever they can to take a piece of what you've built.

It's like being a racehorse in the Kentucky Derby. They put those blinders on the horse so it doesn't get distracted by the crowd, the noise, or the thousands of screaming fans. It only sees the track. That's what you have to do in your life and your business. You have to put blinders on and keep moving forward. Don't let the distractions, the drama, or the people trying to pull you in a million different directions throw you off course. It's about focus. Keep your eyes on your goal and don't let anyone or anything get in the way.

And it's sad, really. Sometimes the people closest to you will be the ones to sabotage you. I've had family steal from me. I had to fire my own sister for it. That's the harsh reality of success. Not everyone will celebrate your wins. Some people will envy you, resent you, or try to tear you down. The people you expect to be your biggest supporters might not be the ones who lift you up. But, as hard as it is, you can't let that get to you. You've got to keep moving forward, knowing that not everyone is going to get it. Not everyone is going to understand the sacrifices you've made or the risks you've taken to get where you are.

The truth is, success will often highlight both the good and the bad in the people around you. It's easy to get caught up in the excitement, thinking that everyone around you will be rooting for you. But the reality is that when you start winning, it makes some people uncomfortable. It forces them to question their own decisions, and that insecurity can breed resentment. It's painful to experience, but it's part of the price you pay for stepping into the spotlight. And when it happens, you have to be ready for it. You can't let it derail your focus or take away from what you've built.

But you know what? You'll also find a few people who stick by you no matter what. My cousin, for example, has been with me since day one. He's built every website I've ever had, managed my e-commerce, and now he's helping me with the app for Outlaw TV. He's my ride or die, and I love him like a brother. Those people—the ones who stick with you through thick and thin—hold on to them. They're the ones who believe in you, even when the road gets tough. It's easy to find fair-weather friends, the ones who show up when things are going well, but it's the people who stand by you when things aren't going so great that really matter. You can't do this alone. It takes a team—whether it's family, friends, or people you've brought in along the way.

You need those loyal people. The ones who will support you when you're at your lowest, when the weight of everything feels too much to bear. Trust me, when things get rough, it's those people who will keep you grounded. They'll remind you of why you started in the first place and help you push through the tough times. You'll need them to protect your vision and keep you focused on your goals when the rest of the world seems intent on pulling you in a million different directions.

At the end of the day, this journey is about you. It's about your skills, your drive, and your belief in what you're doing. Not everyone will understand it. Most people won't. They might question why you're working so hard or why you're so focused on your craft, but if you know in your heart that this is what you're meant to do, you've got to go for it. Ignore the noise. Jump off the cliff. Trust yourself.

Trusting yourself means understanding that your journey is unique. It's about believing that every challenge, every setback, and every victory is part of your path. When you build something from the ground up, it's not just about the end result; it's about everything you learn along the way. Every project teaches you something new. Every failure teaches you resilience. And every success fuels the fire to keep going. It's about the journey, not just the destination. Embrace the lessons, because those are what will shape you and your work in the long run.

That's the heart of it—the drive to keep moving forward, even when it feels like the weight of the world is on your shoulders. But it's important to remember that success isn't a straight line. There are ups and downs. There are moments when you feel like you're on top of the world and moments when you feel like giving up. But if you can keep your eyes on the goal and stay true to your passion, you'll make it. It's not about being perfect. It's about showing up, doing the work, and not giving up. And sometimes, that means knowing when to step away, rest, and come back stronger.

The balance between rest and drive isn't easy to find, but once you figure it out, it's the key to sustaining your passion and protecting your craft. If you push yourself too hard without taking care of yourself, you'll burn out. If you let yourself rest too much, you'll lose momentum. It's about finding that sweet spot where your passion and drive can thrive while also giving your mind and body the space to recharge. That balance is what will protect you, your work, and your future success.

In the end, it's all about staying true to your vision and protecting what you've built. Keep your focus, guard your energy, and surround yourself with the people who truly believe in you. The road ahead might not always be smooth, but with the right mindset and a strong support system, you can navigate any obstacle that comes your way.

21

THE VALIDATION OF RESISTANCE

Now that I'm running Outlaw TV—not tied to any networks, not dealing with their resistance—I've been able to reflect on something. Even back when I was pitching my ideas to networks and fighting to get them made, I realized that there was a certain validation in that resistance. Every time I had a big idea, I had to lean in, dig deep, and sell it—really sell it. I'd tell them how cool it was gonna be, how different it was, and why they should believe in it as much as I did. And when they finally signed off, when they finally wrote the check to make it happen, it was validation.

And here's the thing: I was always right. Every single time, the idea worked. So that was a double validation for me. But now, building my own network and creating shows like *American Craftsman*, I've had to rely on a different kind of validation—my own. There's no one to pitch to anymore. There's no resistance, no gatekeepers. It's just me, my ideas, and my confidence in what I'm doing. That's a different kind of pressure. It's easy to get lost in the "what-ifs"—the doubts, the fears of failure—but I've learned that there's a strength in being your own validation. There's a freedom in not needing anyone else's approval.

When you're starting your own business—whatever it is—you're gonna hit resistance. There are always gonna be pitfalls, setbacks, bumps in the road. But you have to believe in yourself more than anyone else does. You have to trust your ideas, trust your skills, and trust that you'll find a way forward. It's an incredible feeling when it clicks, but it doesn't happen overnight. It takes time. It takes grit. And most importantly, it takes a willingness to keep going, even when it feels like you're just spinning your wheels.

I remember when we first started filming *American Craftsman*. I was really nervous. I'm not afraid to admit that. I had this incredible

A-list production team around me—the best of the best. People I've worked with for years, people I trust. But still, when you're creating something this big on your own, there's a lot at stake. I'll never forget watching the first cut of the first episode. It wasn't even finished—just a rough cut—but I almost teared up. It was that good. The potential for it to suck was really high. You see it all the time—people with a name or a little bit of fame go off and start vanity projects. You know what I mean: "Look at me! I'm so important. I'm so famous. Look at this amazing place I went to!" It's self-serving, ego-driven garbage. Nobody cares. I didn't want this show to be that. I didn't want it to be about me. I wanted it to be real. I wanted it to matter.

And it wasn't just about the content—it was about creating something authentic. For example, we filmed a welding show—something I'm really passionate about—and the first day, I didn't get it. I did enough to pass the test and call it good, but the welds weren't good enough for me. Not visually, not technically. I couldn't show that to people. That's not who I am. So I did what you have to do when you're running your own business: I went back. I paid for another full day out of my own pocket—underwater cameras, full dive team, the whole thing—just to get it right. Just to make it perfect. Because here's the truth: When you're doing your own thing, when you're building something that represents you, you can't settle for "good enough." There's gotta be a switch inside you—an automatic button that flips when you know something isn't right. It's that voice that says: "Do it again. Do it better. Do it right."

22

PUSHING THROUGH MEDIOCRITY

It doesn't matter what you're making—a motorcycle, a car, a gun, a bike, a cabinet, or a TV show. If you're not satisfied, if you know it's not your best work, you've got to go back and push harder. You've got to demand more from yourself. That's what separates the greats from everyone else. Mediocrity is everywhere. Most people are happy just to finish. They'll settle for "good enough" because they're tired, because it's hard, or because they think nobody will notice. But guess what? People do notice. They can see the difference between something that was thrown together and something that was built with pride and discipline. And if you're in a business where what you make has to stand out, where your work has to shine, you can't afford to be lazy about it. That's what separates the people who last from the people who fade away.

I'm not saying you should be a perfectionist—there's no such thing as perfect—but you have to have a standard. You have to have a line you won't cross. It doesn't matter what the project is; if you know it's not your best work, you have to go back and fix it. If you approach everything you do with that mindset—if you push yourself to make sure nothing leaves your hands unless it's perfect—it can't not be successful. I believe that with everything in me. Mediocrity will never bring you success. It may keep you in business for a while, but it won't keep you at the top.

Look, there's no shortage of people who've gone out of business making crap. You know what I mean. Shoddy work, cheap materials, no pride in what they're doing. That's how businesses fail. But I don't know a single person who went out of business making things that were perfect—things that were so good people couldn't help but notice. People can tell when something's built with passion, with skill, and with an eye for detail. They'll recognize the effort, and they'll appreciate it. It's like a bike that has that perfect finish, or a weld that's clean and precise. You can feel it when you look at it. That's the difference between work that stands the test of time and work that gets lost in the crowd.

What I've learned from all of this is that validation is a process, not a one-time event. There's value in the resistance you face when you're starting out. It teaches you to refine your ideas, to become more confident in your vision. But what's more important is that you become your own source of validation. When you're no longer relying on others to tell you that your work is worth something, you learn to trust your instincts, to push through when it feels like no one else is on your side. That kind of self-validation doesn't come easy, but once you get it, it gives you a freedom that nothing else can.

So, what's the takeaway here? Don't settle for mediocrity. Keep pushing yourself to do better, to go further. When you're doing some-thing that matters to you, when you're building something from the ground up, it's on you to make it shine. Don't wait for permission. Don't wait for someone to tell you it's good enough.

Keep pushing until it's perfect. Because when it is, you'll know it. You'll feel it, and so will everyone else. That's the power of doing your best work. It stands out, and it lasts. And most importantly, it validates you—your efforts, your sacrifices, your vision.

23

THE REWARDS OF EXCELLENCE

If you're putting that kind of effort, that kind of discipline, into what you're building, people will come back for more. It doesn't matter what it is. If it's good—really good—it will find its audience. That's the secret to success: consistency. It's about putting in the work every single time, whether it's a big project or a small one. Whether it's something you're doing for yourself or for a client. When you consistently put out your best work, people notice. They appreciate it. And they'll come back to you because they know you don't cut corners.

I've seen this in my own business time and time again. Whether it's the bikes I build or the shows I create, people know when you care about what you're doing. They know when you're not just going through the motions but when you're putting your heart into it. And they'll reward you for it. Whether it's with their business, their trust, or their loyalty, people can sense that level of commitment. They know that you're not going to settle for anything less than the best.

That's the key. If you're going to succeed in anything—whether it's building a business, creating a product, or launching a show—you have to approach it with an unwavering commitment to quality. You have to push through the mediocrity, the easy route, and the shortcuts. You've got to go the extra mile every time. That's the only way you'll build something worth having, something that will last. And that's the kind of work people will remember.

When you're starting your own business— whatever it is—you're gonna hit resistance. There are always gonna be pitfalls, setbacks, bumps in the road. But you have to believe in yourself more than anyone else does. You have to trust your ideas, trust your skills, and trust that you'll find a way forward.

You've got to put your soul into what you make. That's how you build a reputation. That's how you build something that lasts. When people see the effort you put in, when they see you pushing past the mediocrity to make something exceptional, they'll respect it. They'll recognize that you care. That you care enough to make sure your work stands out, to make sure it doesn't just meet the standards but exceeds them.

Look, it's not just about the work you do—it's about the attitude you bring to it. Excellence isn't about being perfect. It's about doing your absolute best, about pushing your limits, about constantly growing and learning. If you're always striving for excellence, if you're always looking for ways to improve, you'll never stop getting better. And that's how you build something meaningful.

The rewards of excellence aren't always immediate. Sometimes, they take time. But when you're committed to doing your best work, when you're committed to being better every single day, those rewards will come. They might not be in the form of instant recognition or fame, but they'll show up in the form of loyal customers, repeat clients, and a reputation that precedes you. People will want to work with you. People will want to support you. Because they know you're the real deal.

And that's the beauty of it. When you consistently produce exceptional work, you don't have to beg for attention. It finds you. It gravitates toward you. Your work speaks for itself. People will notice, and they'll respect you for it. That's the true reward of excellence—not just the financial payoff but the respect and loyalty that come with it. And once you have that, once you've built something that people value and trust, you've got something that will last.

At the end of the day, that's what it's all about. Not just the final product but the process. The journey you take to get there. The commitment you make to always do better, to never settle, to push yourself to create something that matters. That's the kind of work that doesn't fade. That's the kind of work that lasts. And when you're doing that kind of work, you'll know it. Because the rewards will show up, in ways you never expected—and they'll keep showing up as long as you stay committed to excellence.

24

QUALITY WINS IN THE END

Here's the thing: Mediocrity is everywhere. Most people are happy just to finish. They'll settle for "good enough" because they're tired, because it's hard, or because they think nobody will notice. But guess what? People do notice. They can see the difference between something that was thrown together and something that was built with pride and discipline.

I've seen it happen time and time again. People put out something that's rushed and second-rate, thinking that if they just get something out there, they'll be successful. They think that speed will make up for quality. But the truth is, quality will always win in the end. People can feel the difference between something that's just okay and something that's been built with care, precision, and passion. When you take the time to perfect your craft, you send a message—not just to your audience but to yourself—that you're committed to creating something that matters.

If you approach everything you do with that mindset—if you push yourself to make sure nothing leaves your hands unless it's perfect—it can't not be successful. I believe that with everything in me.

And look, there's no shortage of people who've gone out of business making crap. You know what I mean. Shoddy work, cheap materials, no pride in what they're doing. That's how businesses fail. But I don't know a single person who went out of business making things that were perfect—things that were so good people couldn't help but notice. Things that left an impact. You know the businesses I'm talking about. They're the ones that stand out. The ones that leave a legacy because they had the courage to go the extra mile, the strength to push past the mediocrity, and the discipline to make it right every single time.

When you're passionate about what you do, it's hard to cut corners. It's hard to just put something out there without making sure it's the best you can do. And that's where the difference lies. That's what people notice. When you pour your heart into your work, it resonates with others. Your audience can tell when you care. They can tell when you've put in the effort, when you've invested yourself into every detail, every decision.

So, when you're building something—whether it's a YouTube channel, a product, or a service—remember that people can tell the difference between "good enough" and something exceptional. They'll always gravitate toward the exceptional. It's a universal truth. When you're putting that kind of effort, that kind of discipline, into what you're building, people will come back for more. It doesn't matter what it is. If it's good—really good—it will find its audience. You might not always see the payoff immediately, but when people recognize the quality in your work, they'll keep coming back. And when you do it right, they'll bring others along with them. That's how you build something lasting.

But it's not easy. It's never easy. And I'm not saying you should expect perfection every time. There will be mistakes. There will be times when you feel like giving up. There will be setbacks.

But the key is to push through. Don't let the tough times stop you. Don't let the temptation to just "finish" push you toward mediocrity. Keep working until it's right. Keep pushing until you know you've done your best. That's how you turn challenges into opportunities.

And even when you're doing it on your own terms—when there's no network to validate you or gatekeepers to push you—you've got to hold yourself to that same standard. Trust your instincts, your ideas, and the work ethic that got you here. Don't settle for "good enough." Always push for more. Because at the end of the day, it's your name on the work. It's your brand. And your reputation matters more than anything else.

You might not always get immediate feedback, and sometimes the results take longer than expected. But if you continue to hold yourself to the highest standard, that reputation will build. People will take notice. They'll start talking about your work. They'll recommend you. And before you know it, you've got a base of loyal followers, customers, or clients. They won't just come for the product or the service—they'll come because they believe in you. They'll come because they know you deliver something worth their time, something worth their investment.

But success doesn't happen overnight. It's a process. There are no shortcuts. There's no quick way to the top. The road to success is built on a series of small steps, of consistent effort, of pushing through the setbacks and failures. You have to embrace the grind. You have to fall in love with the process of improving, of getting better every single day. When you do that, you create something that lasts. Something that endures.

There will always be distractions. There will always be opportunities to take the easy route. There will always be people telling you to cut corners, to compromise. But the people who succeed, the people who stand out, are the ones who don't listen to those voices. They listen to their own. They trust their own standards. They push themselves, every single day, to be better. They understand that the little details matter. They understand that mediocrity is the enemy of greatness.

And trust me, I've been there. I've faced the temptation to rush things, to settle for "good enough." There have been times when I've been tired, when I've been ready to just get something out there and move on to the next thing. But every time I've given in to that urge, every time I've cut corners, I've regretted it. And every time I've pushed through, every time I've held myself to a higher standard, I've seen the results. The work speaks for itself.

You see, the thing about mediocrity is that it's not just about the work itself. It's about the mindset that drives it. When you settle for mediocrity, you're telling yourself that "good enough" is okay. You're telling yourself that you don't need to push any harder, that you've done enough. But when you push through that temptation, when you refuse to settle, you're telling yourself something different. You're telling yourself that you're capable of more. You're telling yourself that you have the discipline, the focus, and the drive to create something exceptional.

And when you adopt that mindset, when you make it a habit to push through mediocrity every time, it becomes second nature. It becomes part of who you are. You stop seeing obstacles as roadblocks. You start seeing them as opportunities to grow, to improve, and to create something better.

So, whether you're building a business, creating content, or working on a personal project, don't fall into the trap of mediocrity. Push through it. Keep raising the bar. Keep striving for excellence. Because when you do that, you're not just building something that lasts—you're building something that truly matters. And that's what people will remember. That's what will keep them coming back. That's what will set you apart.

25

THE POWER OF SELF-BELIEF

Building Outlaw TV has taught me so much about trusting myself and my instincts. I don't have networks or executives to fall back on anymore. It's all on me. But that's what makes it so rewarding. Every decision I make, every risk I take, it's mine. The success—or the failure—is mine. And I think that's a powerful thing.

Sure, there are challenges. There are days when everything seems like it's crumbling down around you. But even on those days, when it feels like the pressure is too much, you have to remember that you're still in the driver's seat. It's your direction. It's your vision. And that's something no one else can take from you. The weight of ownership can be heavy, but it also gives you the freedom to carve your own path. The roadblocks you hit are yours to overcome—and overcoming them builds something even more important than success: confidence.

Confidence is what carries you through the tough times. It's what makes you wake up every morning and keep pushing, even when it feels like you're swimming against the current. If you don't have that inner belief in yourself, you'll give up at the first sign of trouble. But if you've built your confidence, brick by brick, through every victory and failure, you'll find the strength to keep moving forward.

The beauty of owning your own journey is that you don't have to answer to anyone else. You make the decisions, you set the course, and you're free to explore new paths without fear of someone telling you no. But with that freedom comes responsibility. Every step you take, every choice you make, is on you. That can be intimidating, but it's also incredibly empowering. You realize that you are the one who determines your future. You are the one who gets to define success on your terms.

Self-belief isn't about being cocky or thinking you're invincible. It's about understanding that you are capable of more than you think. It's about trusting your own abilities, even when others doubt you. It's about knowing that you have the strength, the creativity, and the discipline to overcome obstacles and rise to any challenge. And when you have that belief in yourself, when you trust in your own abilities, there is no limit to what you can achieve.

There are always going to be people who tell you it can't be done. There are always going to be critics who want to tear you down. But when you trust yourself, when you believe in your vision and your ability to execute it, those voices become background noise. You tune them out because you know what you're capable of. You know that no one else has the same perspective or the same passion for your idea that you do.

Belief in yourself is the fuel that keeps you going when the road gets tough. When you don't have anyone else to lean on, that self-belief is what carries you through the long nights, the frustrating moments, and the inevitable setbacks. It reminds you that you're capable of more, that every failure is a lesson, and that success isn't defined by how many times you fall but by how many times you get back up.

There's no magic formula to success. There's no secret shortcut. The only way to get there is by doing the work, by pushing through the hard times, and by believing that you're worthy of success. And that belief starts with you. It starts with looking at yourself and saying, "I can do this. I'm going to make it happen."

The truth is, when you trust yourself, you start to notice opportunities everywhere. You start to see solutions instead of problems. Your confidence in yourself becomes a self-fulfilling prophecy—when you believe you can achieve something, you'll naturally find ways to make it happen. You begin to take action, to push through self-doubt, and to reach for your goals.

Self-belief isn't about having all the answers. It's about trusting that you'll find them along the way. It's about knowing that you have the ability to figure it out, even when things seem uncertain. The road ahead may not always be clear, but with confidence in yourself, you can navigate the unknown. And it's through that journey that you discover who you really are and what you're truly capable of.

So, remember this: The power to succeed is already within you. Trust yourself, believe in your abilities, and never stop pushing forward. You're more than capable of achieving everything you set your mind to.

I think a lot of people shy away from taking risks because they're afraid of what might go wrong. But let me tell you something: Risk is part of the game. It's the fuel that drives innovation. Without risk, there's no growth. You don't get better by playing it safe. You get better by stepping out of your comfort zone and trying things that challenge you. It's scary. But I wouldn't have it any other way. Because without that challenge, you don't get to experience the highs that come when you finally succeed.

In fact, that leap is where the magic happens. It's where all the potential lives—the stuff you dream about when you're grinding away, wondering if you're ever going to get where you want to be. It's in those moments of uncertainty that you find out who you really are and what you're capable of. If you're not willing to put yourself out there and take the risk, then you'll never know. You'll never experience the thrill of watching something come to life from nothing. You'll never feel the satisfaction of knowing you did something that no one else dared to do.

And that's a huge part of what makes this whole journey so rewarding. It's the not knowing. It's the chance to push yourself to see what you can really do. To step up to the plate and swing, even though you don't know if you're going to hit a home run. But you try anyway because you believe in the effort. You trust the process. And you trust that you'll learn from the mistakes along the way.

When you take risks, you open yourself up to failure. But here's the thing about failure: It doesn't have to be the end. In fact, failure is one of your best teachers. It's not a reflection of your ability; it's a part of the growth process. You learn what doesn't work, so you can adjust and try again. You refine your approach, you tweak the formula, and you get better. But you can't make those improvements if you're too scared to fail. You can't make progress if you never take that first step into the unknown.

So, take the leap. Bet on yourself. Because, here's the truth: When you believe in yourself, when you back yourself even when it feels like everything is on the line, you'll begin to see the magic unfold. Sure, it's not always going to be smooth sailing. There are going to be bumps in the road. But that's the challenge. And that's what makes it all worth it.

The same principle applies to anything you're working on. Whether you're building a business, creating content, or pursuing a personal goal—if you're not willing to take risks, if you're not willing to push past the fear and step into uncertainty, you're never going to know what you're truly capable of. The first step is always the hardest, but once you take it, you'll find that it gets easier. And with every risk you take, you'll build more confidence, more knowledge, and more resilience.

And this doesn't just apply to the grand leaps. It's about the small daily risks too. The risk of doing something a little bit differently, of stepping outside of your routine, of trying something new that scares you. Those little moments of discomfort, of doing things you're not used to, they build up over time. They shape you. They help you grow. And soon, what once scared you becomes part of your new normal.

But here's the thing: Even when you take risks and push yourself, you can't afford to rest on your laurels. You have to keep showing up. You have to keep taking those risks, keep stepping into the unknown, even when you start to see some success. Success isn't a destination; it's a process. And if you get too comfortable, if you stop taking those risks, you stop growing. You stop evolving. And that's when things start to stagnate.

So whatever you're working on—whether it's a product, a service, a show, or a business—keep pushing. Don't accept mediocrity. Don't cut corners. And don't let yourself off the hook until you know you've given it everything you've got. It's easy to get comfortable, to think that you've made it. But the truth is, you haven't. Not yet. You haven't hit your full potential. And that's what you should be striving for every single day.

Because when you do that—when you show up, push through, and refuse to settle—it's impossible to fail. Sure, there will be setbacks along the way, but those don't define you. What defines you is how you respond. Do you quit? Or do you pick yourself up, dust yourself off, and keep going? The path to success is never linear. It's full of twists, turns, and bumps along the way. But every step you take, every risk you embrace, brings you closer to the person you're meant to become.

It's not about waiting for the perfect moment or the perfect plan. It's about taking action now. Trusting that you're capable of handling whatever comes your way. Learning from your failures, growing from your mistakes, and using them to propel you forward. That's how you build something lasting. That's how you create something that matters.

In the end, it's all about showing up. Every day. No matter what. Keep believing in yourself, keep pushing forward, and you'll find that the success you're seeking isn't as far away as it might seem. Every risk, every challenge, every leap of faith brings you one step closer. And when you get to where you're going, when you finally reach that goal you've been working toward, you'll look back and see that it was all worth it. All the risks, all the sacrifices, all the hard work. Because you didn't just succeed—you grew. You became the kind of person who could handle anything life throws at you. And that, my friends, is the real success.

26

WHY KEEP GOING?

People ask me, "Why do you keep doing this?"

What's the drive? You've hit all these goals, you've done the TV thing, you've got the name, the shop—so what keeps pushing you forward?

I think the truth is . . . I'm just obsessed. Like, I'll be lying in bed at night thinking through how to make a gas tank mount. I'll build the whole thing in my mind—step by step. I'll go through one version, then realize that's not gonna work. So I adjust. Try a different way. Rethink the materials, the welds, the angles. By the time I actually go into the shop, I've already built it in my head a dozen times. That kind of thing doesn't come from chasing success. It's just in me. It's what I'm wired to do.

All the other stuff—meetings, conference calls, visitors, press, even working with big companies like Walmart—it doesn't feel like real work to me. I know it moves the needle and it's important for the business, but I'm constantly thinking, *Okay, how fast can I get through this so I can get back to the real work?*

Even writing this book . . . yeah, I know it's gonna help people, give them insight and perspective they won't get anywhere else—but it still feels like a distraction from the shop. Like, the second I finish this page, I'm already thinking about what I could be machining, welding, fabricating.

A lot of people want to start a shop, build things, make stuff—not because they have to, but because it looks cool. Because they want to be the guy with the shop like Jesse. But me? I couldn't care less about the surface-level shit. I don't see the shop as a whole. I'm zoomed in on micro details. Tiny adjustments. The smallest things that, when they add up, make something incredible. That's where my head's at.

I set micro goals every day. Like, get the front tank mount finished today. Once that's done, then I can move on to the gas cap mount. Then the rear tank mount. Then I'll make a scoop—a little custom ball chiller—for the customer. That's what I focus on. One thing at a time. One obsession after another.

TV? I've done tons of it. *Monster Garage*, series after series, twenty-five years of that stuff. Even when I'm doing it for myself now, not for a network, it still feels like I'm missing something—missing time in the shop. It's better, sure, but it still takes me away from what I was born to do.

Honestly, I think I'm one of the only people like this. There are guys like Mike Rowe—nice dude, met him before—but he's an actor. He visits people who work with their hands, but he doesn't need to. He doesn't have that switch flipped inside him like I do. Me? I have to work. I don't really have a choice.

If all of this went away tomorrow—the shop, the business, the name—I'd still be in a garage somewhere, working with my hands. Because I have to. That's what keeps me going.

CONCLUSION

I'll never forget that letter I mentioned earlier in this book, from that fifteen-year-old girl in Illinois. After *Monster Garage* became a hit, she wrote to me saying, "Thank you for making my dad respectable." Her father was a diesel mechanic, and like most mechanics, he'd come home every night in grease-covered overalls, smelling of oil and exhaust. People look at jobs like his—welders, fabricators, mechanics—as "dirty work," almost shameful. But to her, seeing someone like me doing what her dad did made her realize the respect he deserved.

That letter hit me harder than any other I've received. It stuck with me because that's the problem: Society has dumbed down how it sees tradesmen, mechanics, and craftsmen. People who work with their hands are often viewed as "grease monkeys" or "primitive apes," as if these jobs require no skill or intelligence. But that's a lie. These jobs—whether you're welding, machining, fabricating—require incredible talent. The guys I know who do this work at a world-class level are geniuses. They can figure out how to make a machine or build a structure from nothing, using their hands, their minds, and a few simple tools. The problem is that we've become too enamored with the idea of a "clean desk job" to recognize the importance of manual labor.

This book is for those who respect that difference—the people who want to bring real craftsmanship back to their lives. It doesn't matter if you've never picked up a hammer or touched a forge before. If you're willing to learn, to get dirty, and to fail along the way, you can build something incredible. Whether you're an office worker, a weekend warrior, or someone dreaming of a different life, the message is simple:

Respect the trades and the value of working with your hands.

Because at the end of the day, real life is where it matters—not YouTube.

Start learning.

Start building.

Start creating something real.

ABOUT THE AUTHOR

Jesse James is a master fabricator, entrepreneur, and lifelong builder who forged his way from a turbulent childhood in South Central Los Angeles California to become one of the world's most iconic craftsmen. Raised around his father's furniture restoration shop and a neighbor's motorcycle machine shop, Jesse learned early that hard work, grit, and precision were the price of pride. He didn't choose the path of traditional education—instead, he found purpose in working with his hands, building his first custom bikes in a two-car garage, and grinding through every setback on his own terms.

As the founder of West Coast Choppers and Jesse James Firearms Unlimited, Jesse built not just a business, but a legacy—one that's powered by an unrelenting drive to make things that matter. Every piece he creates—whether a motorcycle, custom car, firearm, knife, or heirloom cookware—is rooted in discipline, respect for craft, and a belief that fulfillment comes from the struggle, not shortcuts.

Now based in Austin, Texas, Jesse continues to build every single day—not for fame or recognition, but for the satisfaction of knowing he's creating something real, and providing customers more than they are expecting. His work is a reflection of who he is: relentless, hands-on, and deeply connected to the process. Always striving to be the best.